Saint Peter's University Library
Withdrawn

Saint Peter's University Library
Withdrawn

Eukaryotic Cell Genetics

This is a volume in
CELL BIOLOGY
A series of monographs

Editors: D. E. Buetow, I. L. Cameron, G. M. Padilla, and A. M. Zimmerman

A complete list of the books in this series appears at the end of the volume.

Eukaryotic Cell Genetics

JOHN MORROW

Department of Biochemistry
Texas Tech University School of Medicine
Lubbock, Texas

1983

ACADEMIC PRESS

A Subsidiary of Harcourt Brace Jovanovich, Publishers

New York   London

Paris   San Diego   San Francisco   São Paulo   Sydney   Tokyo   Toronto

COPYRIGHT © 1983, BY ACADEMIC PRESS, INC.
ALL RIGHTS RESERVED.
NO PART OF THIS PUBLICATION MAY BE REPRODUCED OR
TRANSMITTED IN ANY FORM OR BY ANY MEANS, ELECTRONIC
OR MECHANICAL, INCLUDING PHOTOCOPY, RECORDING, OR ANY
INFORMATION STORAGE AND RETRIEVAL SYSTEM, WITHOUT
PERMISSION IN WRITING FROM THE PUBLISHER.

ACADEMIC PRESS, INC.
111 Fifth Avenue, New York, New York 10003

United Kingdom Edition published by
ACADEMIC PRESS, INC. (LONDON) LTD.
24/28 Oval Road, London NW1 7DX

Library of Congress Cataloging in Publication Data

Morrow, John, Date
   Eukaryotic cell genetics.

   (Cell biology)
   Bibliography: p.
   Includes index.
   1. Cytogenetics.   2. Eukaryotic cells.   I. Title.
II. Series.
QH430.M67   1982   599'.08762   82-11608
ISBN 0-12-507360-7

PRINTED IN THE UNITED STATES OF AMERICA

83 84 85 86   9 8 7 6 5 4 3 2 1

QH
430
M67
1983

CONTENTS

PREFACE

This book represents an effort to accomplish several tasks. Its main purpose is to describe in a brief and topical fashion the present state of knowledge of somatic cell genetics. This implies the predominant use of certain techniques, mainly the use of normal and transformed mammalian cells grown *in vitro* and their genetic analysis through the techniques of molecular biology, biochemistry, cytogenetics, and cell hybridization. Although much of what this volume deals with is precisely this, it is necessary to go beyond these approaches, especially in terms of the broad interpretation of these findings. All of the fundamental processes which we wish to consider are the result of genetic structures transmitted through a sexual cycle and expressed in somatic cells in a particular fashion. Thus descriptions of particular findings at the cellular or molecular level have frequently been presented in relation to medical, human, and population genetics.

I have not attempted to cover the various topics in a historical context nor to review them completely. There are a number of excellent books which have reviewed the older literature in cell heredity and cell culture, and the reader is referred to them throughout for the foundation of particular areas. Similarly, I have included a number of unresolved or controversial topics which, because this area is moving with such rapidity, may be resolved by the time this book is published.

I have aimed this book at an audience which includes researchers in the fields of genetics and molecular biology, nonspecialists interested in what is happening in a very exciting area of biology, and students at the graduate level in cell biology. I was forced to make compromises between highly detailed explanations of particular experiments and broader outlines which attempt to integrate findings in a particular field. In a number of instances, not all of the examples of a given phenomenon are listed, but representative cases are described.

Several friends, colleagues, and students gave me valuable comments and criticisms. These include David Patterson, Abraham Hsie, Milton Taylor, Olga Zownir, Stanislaw Cebrat, Daniel Meier, David Hong, Phillip Keller, Susan Jones, Donald Clive, Court Saunders, and Jean Orme. Their advice has been extremely helpful in the development of this volume. The original figures were most competently drawn by Harvey Olney. Finally, I wish to express my thanks to Shirley Gaddis for her invaluable assistance in assembling references, proofing, and typing the manuscript.

John Morrow

Eukaryotic Cell Genetics

1

Somatic Cell Genetics and the Legacy of Microbial Systems

I. INTRODUCTION

In the last 20 years our understanding of the genetics of higher organisms has undergone tremendous growth and metamorphosis. This is not surprising because during this period all of biological science has expanded tremendously and our level of comprehension has thereby increased. However, the genetics of eukaryotic cells is particularly striking in this regard. This progress resulted from the application of techniques found to be successful in bacterial genetics and molecular biology to the cells of differentiated, multicellular organisms. Because of the practical implications of these findings for the biomedical sciences, results in somatic cell genetics have been greeted with great interest and enthusiasm and research on the genetics of higher cells has been well supported by the federal government, as well as by private foundations.

The development of somatic cell genetics has to a large extent concerned itself with two major needs: a search for variability in cell populations, and an investigation of mechanisms of mating or of exchanging genetic information between cells. These directions are, of course, readily understandable because isolated somatic cells do not represent real organisms in a biological sense but rather derive in an artificial manner from whole individuals that are broken into disaggregated collections of cells which are treated as microbial populations. Now that the problem of satisfying those two basic necessities for genetic analysis is largely resolved, the study of heredity in somatic cells concerns itself with the architecture of genetic elements in eukaryotes, including their interaction through regulative function. In the process of developing our understanding of these questions, fundamental processes common to higher organisms, including

1

aging, immunology, differentiation, and neoplasia are gradually becoming understood.

Because the technical methodology of cell genetics resembles that which has been so widely utilized in microbial genetics, it is not surprising that the rationale and approaches would also be derived from similar sources. The idea of applying the techniques and experimental approaches of bacterial genetics to eukaryotic somatic cells represented a great asset in that it lent itself effectively to quantitative studies that formed the basis for a later thorough understanding of mammalian genetics and molecular biology. At this time, a detailed historical analysis is probably not warranted. However, one basic approach deserves consideration because of its widespread use in cell heredity and because it represents a fundamental link between microbial and eukaryotic cell genetics.

II. MUTATION VERSUS ADAPTATION IN BACTERIAL POPULATIONS

Since the nineteenth century, it has been known that bacterial populations are not completely homogeneous. Later, variation within a population was shown in terms of nutritional requirements, drug and virus resistances, cell wall characteristics, and pigment production. The basis of this variation, although genetic, was not well understood because no experimental means existed whereby the hypothesis of selection could be distinguished from that of adaptation. Thus, when a population of bacteria was plated in the presence of bacterial virus, rare resistant clones arose; whether these clones were preexisting in the population or induced by the presence of the virus was a question with important philosophical and experimental implications. If variants were occurring all the time through spontaneous mutations and were selected out through intervention of the selection agent, then this would indicate that bacteria possessed genetic systems comparable to those in higher organisms which were known to be subject to the rules of classical Darwinian evolution and natural selection. If, on the other hand, the variants were induced by the presence of the virus, this would suggest that such populations might be quite different genetically from higher organisms because this would argue for the existence of a directed mutational process.

Distinguishing between these two possibilities proved to be extremely difficult, because it is not at all obvious in what way one could test for the existence of variants without exposing them to the agent in question. The problem was solved in a most elegant manner through an indirect mathematical approach. The "fluctuation test" of Luria and Delbrück (1943) consists of two components: first, a control series in which a larger number of cells are plated on a large number of plates in an identical fashion in the presence of the selecting agent in order to calculate a mutation frequency. The distribution of numbers of variants from culture to culture should be random and should follow the Poisson distribution.

Second and parallel to this, there is an experimental series in which secondary cultures that were initially grown from small samples of cells of the primary population are harvested independently and plated in the selecting agent. In the experimental series, the Poisson distribution would also be observed if the mutants originated at the time of the addition of the selecting agent, because the only source of variability in their numbers would be the statistical fluctuation. On the other hand, if the variants were preexisting in the primary population, then each independent experimental culture would reflect a slightly different history: in some, mutants would have arisen early in the growth period; in some, later; and so forth. Thus, the variance of the distribution of mutants in the secondary populations would be much greater than in the control and would not follow the Poisson equation (Fig. 1.1). It follows from this that a difference in the variation

BOTTLE OF CONFLUENT TISSUE CULTURE CELLS

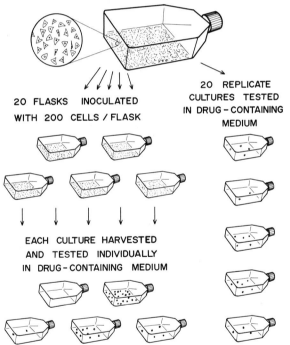

20 FLASKS INOCULATED
WITH 200 CELLS / FLASK

20 REPLICATE
CULTURES TESTED
IN DRUG – CONTAINING
MEDIUM

EACH CULTURE HARVESTED
AND TESTED INDIVIDUALLY
IN DRUG – CONTAINING MEDIUM

Fig. 1.1. Protocol of the fluctuation test as adapted to cultured cells. A bottle of confluent cells is harvested, and large numbers (approximately 10^6 cells) are plated in individual flasks as shown in the right-hand column. This constitutes the control series. At the same time a series of flasks are initiated from small starting numbers of cells from the primary bottle and are grown in normal medium. These are experimental cultures that are then harvested and plated in the selecting agent. Because each experimental culture has a slightly different history, the frequency of mutations will vary much more widely from flask to flask in the experimental series than in the control series.

("fluctuation") between the control series and the experimental series would indicate that the variants were preexisting in the primary population and that the toxic agent merely eliminated the wild-type members from the culture.

The fluctuation test has been widely utilized with a number of systems, including bacteria, mammalian cells, yeast, and other microorganisms. In almost every case the hypothesis of selection rather than adaptation has been proved. This point is so well established that it is taken for granted and would not merit such extensive consideration in this volume were it not for the fact that the fluctuation test also allows one to measure the mutation rate for the particular variant under selection.

Because the experimental series is started from extremely small numbers of cells, all preexisting mutants have been eliminated through dilution. Because the number of cells plated and the number of mutants obtained are measurable quantities, a rate can be obtained through application of appropriate equations. The average number of mutants yielded per culture is calculated through methods derived by Luria and Delbrück (1943) and later expanded upon by Lea and Coulson (1949) and Capizzi and Jameson (1973). A discussion of these methods and some examples of their application are given in Lea and Coulson, (1949). The three methods used for calculating the average number of mutants per sample and the mutation rates are (1) P_0, (2) median, and (3) maximum likelihood. Use of these methods depends upon the distribution of numbers and the degree of accuracy required by the investigator. Usually close agreement is obtained among the different methods (Table 1.1).

In any population of microorganisms, mutations are occurring and might be expected to eventually take over the entire population. The fact that this does not occur (unless, of course, selective pressures change) demonstrates that new mutants are at a selective disadvantage and are eliminated from the sample. By this mechanism, an equilibrium level will be established in which the rate of addition of new mutants to the population through mutation will be exactly counterbalanced by their elimination through negative selection. The reason for this selective disadvantage is not immediately obvious, but no doubt arises from the fact that most mutations represent a disruption in the normal, finely tuned system and thus would not be expected to compete effectively with the wild type. Under these conditions, the mutation frequency and the mutation rate will not be the same.

Equations have been derived (Morrow, 1964) that give the relationship of the mutation frequency to the mutation rate. Let μ = forward mutation rate, m = number of mutants, n = number of wild-type cells, s = generation time of mutant/generation time of wild type, F_0 = fraction of mutants at generation zero, and F_t = fraction of mutants at generation t. At any generation, F_t will be the result of two factors: the contribution due to the initial frequency F_0 times s to the power t, and the contribution due to mutation, summed for t generations and

TABLE 1.1   Example of Mutation Rates Obtained Using the Fluctuation Test[a] for the Mutational Step from Asparagine Requirement to Independence[b]

| | Number of cultures | | | | | | |
| | Polyploid experiment | | | Control experiments | | | |
	1	2	Pooled data	1	2	3	Pooled data
Number of clones							
0	11	11	22	4	3	0	7
1	7	4	11	2	3	4	9
2		2	2	2	0	3	5
3		1	1	2	2	1	5
4				1	0	1	2
5	1		1	0	0	2	2
6				2	1	0	3
7				1	0	1	2
8		1	1	0	3	1	4
10				2	1		3
11				1	1		2
13					1		1
16							
17	1		1	1		1	2
19					1		1
104						1	1
Number of starting cells/culture	10^3	10^3	10^3	10^3	10^3	10^3	10^3
Cells plated in Asn medium $\times 10^6$	0.96	1.2	1.08	1.7	1.5	2.0	1.7
Mean number of revertant clones/flask	1.3	1.4	1.4	2.5	2.6	2.0	2.45
P_0[c]	0.550	0.579	0.564	0.222	0.188	0.000	0.143
Mutation rate							
Method 2[d]	1.4	1.2	1.3	1.5	1.7	1.0	1.4
Method 1[e]	0.6	0.47	0.53	0.90	1.1	—	1.1
95% confidence limits on pooled data, Method 1			0.15–0.96				0.71–1.7

[a] From Prickett et al. (1975).

[b] Numbers of clones obtained by plating cells in asparagine-free medium and after counting numbers of macroscopically visible colonies for 2 weeks.

[c] P_0 = Number of flasks with 0 clones/total number of flasks.

[d] Median method.

[e] P_0 method.

adjusted for the growth rate differential. Because the frequency of mutants is always small, the effect of reverse mutation may be ignored. Therefore

$$F_t = \mu(1 + s + s^2 + s^3 + s^4 + \ldots + s^{t-1}) + s^t F_0 \qquad (1.1)$$

which may be expressed as

$$F_t = \mu\{[1 + s/(s-1)](s^{t-1} - 1)\} + s^t F_0 \qquad (1.2)$$

At equilibrium the second term will vanish and the equation will reach its limiting form

$$m/n = \mu/(1-s) \qquad (1.3)$$

We have published the results of fluctuation test studies using V79 hamster cells in which both the frequency and mutation rates were calculated (Morrow *et al.*, 1978). From these data an s value of 0.94 for the mutant with respect to the wild type can be calculated indicating the mutant cells are at a slight selective disadvantage when grown together with wild type. This estimate agrees well with that obtained a few years ago (Morrow, 1972) on the basis of artificial mixtures of drug-resistant and drug-sensitive permanent mouse cells. Thus, it appears that we can account for the population dynamics of mammalian cells on the basis of a simple mutational model: such cells represent a population in which new variants are constantly arising through mutation and being eliminated through natural selection. It does not appear necessary to develop more complex mathematical models of mutation and selection in cultured mammalian cells to explain these observations.

III. THE BASIS OF VARIATION IN SOMATIC CELLS

Although their molecular basis is not completely understood, mutation rates obtained by the fluctuation test have been a frequent topic of investigation, and this problem is one of both basic and practical interest. As shown in Table 1.2, the fluctuation test has been widely applied to a variety of cellular systems. The precise nature of these variants will be considered in detail in subsequent chapters, but there are several generalizations that merit emphasis. In the first place, there does not appear to be a difference in mutation rates between dominant and recessive mutations (Table 1.3). Because the cells in question are presumably at least diploid, this is a difficult observation to understand. This concept is illustrated in Fig. 1.2 for a typical diploid locus. In order for a dominant mutation to be expressed, only one mutational event is required, occurring with a rate x. However, a recessive mutation requires two simultaneous events with a combined probability of y^2. Because x and y are in the same range ($x \simeq y$), x should

TABLE 1.2 Mutation Rates Compiled from Fluctuation Test Data for Mammalian Cells in Culture[a]

Character	Mutation rate $\times 10^{-6}$	Cell line	Comments	Reference
HPRT loss	1.5–5.7	Heteroploid mouse cells	Recessive, X-linked hemizygous	Morrow (1970)
	0.45–1.8	Human diploid fibroblasts		DeMars (1974)
	15	Established hamster		Shapiro et al. (1972)
	0.018	Established hamster		Chu et al. (1969)
	22	Established hamster		M. Harris (1971)
	70	L-54 diploid human (male)		Shapiro et al. (1972)
	70	Transformed W1-38 (human)		Shapiro et al. (1972)
	1	D98/AG (HeLa)		Szybalski and Szbalska (1962)
	3.1	Established hamster		Morrow et al. (1978)
TK loss	0.12	Mouse lymphoma	Assumed $+/- \rightarrow -/-$	Clive et al. (1972)
	<0.035	BHK established hamster	Recessive	Caboche (1974)
Ouabain resistance	0.06	Chinese hamster	Dominant	Baker et al. (1974)
Loss of heavy or light chain production	1100	Mouse myeloma	Possibly related to immunoglobulin variation	Coffino and Scharff (1971)
Diphtheria toxin resistance	15	Human diploid fibroblasts		Gupta and Siminovitch (1978)

(continued)

TABLE 1.2—Continued

Character	Mutation rate $\times 10^{-6}$	Cell line	Comments	Reference
Dibutyryl cAMP resistance	0.2	Mouse lymphoma		Coffino et al. (1975)
Asparagine independence	0.71–1.7	Jensen sarcoma	Dominant, may be revertant	Prickett et al. (1975)
	0.74–0.9	L5178Y mouse lymphoma		Summers and Hand-schumacher (1973)
	1.4–3.5	Walker carcinosarcoma		Morrow (1971)
Loss of HLA antigen	0.78	Diploid human lymphoid line		Pious and Soderland (1977)
Benzo[a]pyrene resistance	0.2	Mouse Hepa-1	Differentiated trait	Hankinson (1979)
Phosphonacetyl-L-aspartate resistance	22	SV40-transformed syrian hamster cell		Kempe et al. (1976)
Puromycin resistance				
First step	4	Mouse L cells		Lieberman and Ove (1959)
Second step	100			
	0.1	Hamster cells V79	Dominant	Morrow et al. (1980)

8

Characteristic	Concentration	Cell type	Genetics	Reference
Glucocorticoid resistance	3.5	Mouse lymphoma pseudodiploid	Recessive (?)	Sibley and Tomkins (1974)
Phytohemagglutinin resistance	1.5	CHO	Recessive	Stanley et al. (1975b)
Aminopterin resistance	10	Murine lymphoblastoma	Codominant	Mathis and Fischer (1962)
Low glucose tolerance	100	Established hamster	May be multiple loci	Shapiro et al. (1972)
2-Deoxygalactose resistance	10–35	Established hamster	Recessive	Claverie et al. (1979)
Emetine resistance	0.49	Established hamster	Recessive; hemizygous (?)	Gupta and Siminovitch (1976)
Diaminopurine resistance	0.3	Established hamster	Possible $+/- \rightarrow -/-$	Jones and Sargent (1974)
Tubercidin resistance (adenosine kinase deficiency)	10	Established hamster (CHO)	Recessive	Rabin and Gottesman (1979)
Cyclohexamide resistance	10	CHO		Pöche et al. (1975)

[a] Modified from Morrow (1975).

9

TABLE 1.3   **Average Mutation Rates for Characters in Table 1.2 Whose Dominance–Recessivity Relationships Have Been Analyzed in Cell Hybrids**[a]

Character	Mutation rate
Recessive markers[b]	6.9×10^{-7}
Dominant markers	1.9×10^{-7}
Hemizygous (X-Linked; HPRT only)	4.2×10^{-6}

[a] Modified from Morrow (1975).
[b] Averages were obtained assuming a log-normal distribution of mutation rates and are expressed per cell per locus per generation.

be much greater than y^2. In the second place, although the average mutation rate for all the characters considered is not excessively high (by the standards of microbial genetics), it appears to be higher than would be expected purely on the basis of extrapolation of *in vivo* mutation rates in man. This *in vivo* figure (for an unbiased estimate of sex-linked recessives in man) is 3×10^{-7} (Cavalli-Sforza

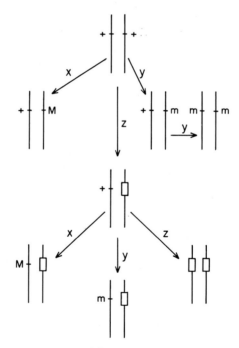

Fig. 1.2.   Schematic representation of three types of variation in cultured cells. $+/+$, A diploid, wild-type locus; x, rate of dominant mutation; y, rate of recessive mutation; z, rate of gene inactivation. zx and zy constitute the model of somatic variation proposed by Siminovitch (1976). zx, zy, and zz constitute the steps of the model proposed by Morrow (1977).

and Bodmer, 1971). Because there are approximately 100 to 1000 cell divisions between each gametic union (in males), this would suggest that the average mutation rate should be of the order of 10^{-9} per locus per generation (Morrow, 1975). But cellular mutation rates are higher than this by two to three orders of magnitude, suggesting that cultured cells are particularly unstable from a genetic standpoint. Finally, normal diploid fibroblasts do not appear to have higher rates of mutation than do highly malignant cells [Table 1.2; hypoxanthine guanine phosphoribosyltransferase (HPRT) mutation rate].

The cause of elevated *in vitro* mutation rates could arise from several factors. Siminovitch has proposed (1976) and has offered experimental support (Gupta *et al.*, 1978a,b) for the hypothesis than some permanent cell lines are functionally haploid for large stretches of the genome, possibly as the result of gene inactivation. These experiments will be considered in detail in later chapters, but the basis for such gene inactivation could reside in chromosomal rearrangements and position effects causing gene inactivation similar to that observed in classic experiments in *Drosophila* (Baker, 1968). If many genes were already functionally haploid because of gene inactivation (z; Fig. 1.2), then recessive mutations would be much more easily realized in culture (zy; Fig. 1.2). This would also explain the observation that dominant and recessive mutations have similar rates (zx; zy; Fig. 1.2). The possibility that the Chinese hamster ovary (CHO) cell line has large blocks of hemizygous genetic material has been tested by Siciliano *et al.* (1978), who isolated by nonselective means 10 mutants affected in their electrophoretic patterns for various "housekeeping" enzymes. Only one of these gives a pattern of bands indicative of hemizygosity (i.e., the presence of only one functional allele), thus failing to support Siminovitch's (1976) model of extensive functional hemizygosity in the CHO cell line.

However, gene inactivation could be an important factor in the elevation of mutation rates *in vitro* if both alleles were subject to inactivation at a high rate so as to bring about the loss of a particular function. Such gene inactivation, combined with point mutation and other chromosomal changes, could explain to a large extent the high mutation rates in cultured cells (zz, zx, and zy; Fig. 1.2).

A paper by Bradley (1979) provides some evidence for the importance of gene inactivation in mammalian cells. He has observed that in a presumptive heterozygote for thymidine kinase (see Chapter 2) the gene could be inactivated and reactivated with a high frequency. Furthermore, a spreading effect down the length of the chromosome appeared to exist because his evidence suggested that the closely linked galactokinase gene could be simultaneously affected.

There are other explanations for the discrepancy between mutation rates *in vitro* and *in vivo*. It is quite probable that the process of growing cells *in vitro* introduces a plethora of instabilities. For instance, it has been shown that visible light is mutagenic (Jostes *et al.*, 1977); furthermore, tissue culture medium contains a variety of compounds that are potentially capable of elevating the mutation rate (purines, pyrimidines, phenol red, antibiotics).

The bias in the selection of cell types is also an important consideration. Because in some cell lines it is extremely difficult to obtain mutants, investigators have gravitated toward lines (such as CHO) in which mutants can be obtained with ease. Thus, the mutation rates reported in somatic cells may not represent a random sampling of all mutation rates but may be biased toward more unstable loci and cell lines.

Furthermore, several technical features of the reported data might tend to artificially elevate the mutation rate. For example, in many studies it is not clear whether every colony of cells appearing in the presence of the selecting agent is in fact genetically resistant rather than simply altered physiologically so that it can complete a few generations and give rise to a macroscopically visible colony. Because in many cases a random sample of colonies has not been collected and tested for permanence of the mutant state, it is not clear to what extent mutation rates might be artificially exaggerated through the inclusion of nonmutant colonies. In those cases in which such controls have been performed, many of the clones obtained appear not to be true resistant variants (Carson et al., 1974).

Fox and Radacic (1978) have published an extensive analysis of low-level purine analog resistance. This study indicates that at least some of the clones arising at low drug concentrations are of nonmutational origin and may represent adaptational changes to the drug employed in their selection. Further selection results in a stabilization and the appearance of the familiar stable types of variants that appear to be true gene mutations (Chapter 2). Removal of the partially resistant lines from the selecting agent results in a gradual erosion of their resistance over a period of 6 to 9 weeks and a return to wild-type behavior. In the case of methotrexate resistance, certain resistant clones also display a rapid erosion of resistance when removed from the selecting agent (Chapter 2; Schimke, 1980). This decline appears to be due to the loss of minute chromosomes that lack a centromere and that carry the DNA sequence responsible for resistance. This phenomenon may be widespread in cultured cells and could explain much of the instability of drug-resistant phenotypes.

It has been suggested on numerous occasions that somatic recombination might be an important factor in the generation of variability in cultured somatic cells and that it might explain such features as high mutation rates, occurrence of recessive mutants in putatively diploid cell lines, and instability or reversion of drug-resistance markers. Rosenstraus and Chasin (1978) have published an extensive analysis of the separation of linked markers in cell hybrids, and their results indicate that if somatic recombination is occurring its rate is less than 10^{-6} per cell per generation. This report, based on the behavior of segregants in clones doubly heterozygous for glucose-6-phosphate dehydrogenase (G6PD) deficiency and HPRT deficiency (azaguanine and thioguanine resistance) would argue that somatic recombination makes little, if any, contribution to the production of altered cell types.

Another possible source of variability has received substantial scrutiny in recent years. It has been recognized that gene expression in eukaryotic organisms can be conditioned by DNA methylation patterns (Burdon and Adams, 1980) and that such alterations can permanently be fixed in progeny cells. When DNA is methylated, especially in the cytosine residues of CpG sequences, this pattern on the parent strand can be replicated semiconservatively through the action of a maintenance methylase. Such methylated sequences undertranscribe messages and many active gene sequences may be hypomethylated when compared to the total cellular genome. Harris (1982) has provided evidence that certain thymidine kinase negative cell lines may result from DNA methylation since exposure of such cells to agents that induce hypomethylation results in massive reversion. Methylation has been suggested as having a role in X chromosomal inactivation, regulation of differential gene expression, and control of drug resistance. Although the ubiquity of this phenomenon cannot be decided at present, any discussion of cell variability must take the possibility of hypomethylation into account.

We therefore appear to be left in a quandary. Although the genetic basis of a number of variants has been resolved at this time (Chapters 2 and 3), we still appear to lack an understanding of the "average" cell culture variant; whether the most common mechanism of variation in cultured cells is gene mutation, chromosome loss, somatic crossing over, gene amplification, or some sort of modulation remains for the future to decide.

IV. SUMMARY AND CONCLUSIONS

Somatic cell genetics initially derived its theoretical and experimental orientation from microbial genetics. In its early stages the fluctuation test was adapted to cell culture systems and a variety of mutation rates were calculated. These studies established that every character appeared to be preexisting in the population rather than induced by the selecting agent, suggesting that somatic cell genetics would prove to be an extension of microbial genetics. Mutation rates calculated from fluctuation test data have yielded numbers that on the average appear to be higher than that which would be expected on the basis of extrapolation of *in vivo* mutation rates in man. Although there are some technical and artifactual explanations for this phenomenon, the overall high rates do tend to indicate that mammalian cells may be more unstable *in vitro* than *in vivo*, possibly due to several factors, including a high frequency of chromosomal rearrangements, detachment of genetic information, and position effects.

Some recent studies indicate that drug-resistant variants selected at low concentrations may be unstable and may not represent true gene mutations. Their relationship to true gene mutations (to be discussed in subsequent chapters) remains to be clarified.

Drug Resistance and Its
Genetic Basis

I. INTRODUCTION

In Chapter 1 we discussed the estimation of mutation rates in mammalian cells, with frequent reference to the use of drug resistance as a selective character. In this chapter we will consider some of the specific properties of such variants in an effort to understand their molecular basis. In a historical sense, it is appropriate to first consider the chemotherapeutic agents. They were originally utilized in cancer treatment, and the study of drug resistance in mammalian cells grew out of an effort to understand drug-refractory states (Harris, 1964). Thus, it was a common observation that treatment of human experimental neoplasms with anticancer agents resulted in the emergence of resistant lines which no longer responded to the agent in question.

The fluctuation test was used for the first time in a mammalian system (Law, 1952) to demonstrate that amethopterin- (see Fig. 3.5) resistant variants were preexisting rather than induced by the selecting agent. This experiment employed variance in the weight of tumors derived from separate tumor substrains as compared to replicates of the same substrain in animals that had been treated chemotherapeutically with amethopterin. Subsequent to this, a large body of work was initiated on the purine analogs, yielding our most detailed understanding of any eukaryotic cellular variant.

II. PURINE ANALOGS

Purine analogs proved to be highly useful selective agents, in part because the natural purines are not essential metabolites for mammalian cells and thus can be

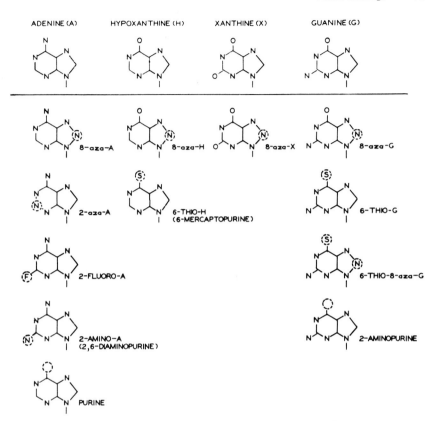

Fig. 2.1. Structure of some of the common purines and their analogs. (From Szybalski and Szybalska, 1962.)

eliminated from tissue culture media. The most commonly employed purine analogs are azaguanine, thioguanine, and diaminopurine (DAP) (Fig. 2.1).

A. Azaguanine and Thioguanine

Azaguanine, an analog of hypoxanthine, was employed in pioneering studies by Szybalski and Smith (1959). Subsequently, thioguanine, a somewhat more satisfactory selecting agent, was used in numerous investigations. Figure 2.2 shows a broad outline of mammalian purine synthesis and the critical enzymes involved.

The toxicity of the purine analogs is not entirely understood; however, it appears that azaguanine interferes with the formation of guanosine monophosphate (GMP) and possibly may be involved in faulty messenger synthesis, while thioguanine appears to exert its effect through incorporation into DNA (Nelson *et*

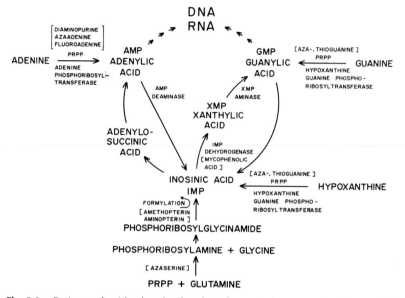

Fig. 2.2. Purine nucleotide phosphoribosyltransferases in humans and pathways of inter-conversion of normal purines.

al., 1975). In any case, plating of cells in increasing levels of the analogs results in a decline in cloning efficiency at inhibitory concentrations. Above the inhibitory dose, it is possible to select for rare resistant variants (Fig. 2.3). By subsequent increments in the dose of the analog, it can be demonstrated that several levels of resistance exist in mouse (Morrow, 1970; Littlefield, 1964), human (Szybalski and Szybalska, 1962), and hamster (Fox and Radacic, 1978) cells. In the mouse, at least three levels of resistance can be detected, and the degree of resistance is inversely related to the level of HPRT (Table 2.1). Such enzyme-deficient cells escape the toxic effects of fraudulent nucleotide synthesis. In these experiments it is necessary to use dialyzed fetal calf serum because hypoxanthine present in undialyzed serum can confuse the results (Peterson *et al.*, 1976).

Our understanding of the genetics of HPRT deficiency in cultured cells was given a substantial boost by the finding of Seegmiller *et al.* (1967) that a loss of this enzyme was responsible for a rare, bizarre, sex-linked defect, the Lesch–Nyhan (LN) syndrome. Affected males with this fatal disorder are characterized by compulsive self-mutilation, choreoathetosis, developmental retardation, spasticity, and an accelerated rate of purine biosynthesis (Caskey and Kruh, 1979). Subsequently, it was determined that the structural gene for HPRT was located on the human X chromosome. The Lyon hypothesis (Chapter 13) predicts that mothers of such individuals (the former being, perforce, heterozygous)

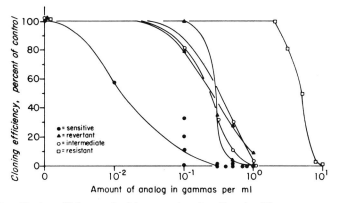

Fig. 2.3. Cloning efficiency of wild-type and variant lines in different concentrations of azaguanine (Morrow, 1970).

should be a composite of positive and negative cells. This prediction is based upon the fact that mammalian females possess only a single functional X chromosome in each cell, and the functional X may, on a random basis, be either of paternal or maternal origin. Such a mechanism is necessary to ensure that both males and females will produce the same amount of X-linked gene products (Lyon, 1972). In fact, there appears to be some selection, and the blood cells of such individuals have a preponderance of HPRT$^+$ cells, possibly due to a selec-

TABLE 2.1   Specific Activity (in Counts of IMP Minute^{-1} μg Protein^{-1}) of HPRT among Derivatives of Wild-Type Cells Grown under Various Conditions[a]

			Specific activity		
Cell line	Selecting agent	Level of resistance[b]	Low	Intermediate	High
M-Mc	. . .	0			17.7, 14.8
Mc-3	. . .	0			11.6
McTG^{R16}	0.005	ND[c]		4.4	
McTG^{R18}	0.010	ND		6.4	
McTG^{R19}	0.080	ND		3.8	
McTG^{R22}	0.100	0.10		3.2, 6.7, 5.5	
McTG^{R25}	0.200	ND		3.3	
McTG^{R17}	0.300	1.0	0.4		
McTG^{R20}	0.500	1.0	0.1		
McTG^{R2}	1.000	1.0	0.1		
McTG^{R6}	1.000	1.0	0.6		

[a] From Morrow (1970).

[b] The second column gives the amount of azaguanine in micrograms per milliliter in which the cell line was selected.

[c] ND, Not determined.

tive advantage of the enzyme-positive cells. This may arise from a need of the erythrocytes for the purine salvage pathway in order to answer their metabolic demands.

It is now clear that the genetic defect in LN syndrome and at least some of the mutational changes responsible for purine analog resistance have the same basis, as proved by the following considerations:

1. Both involve mutations in structural genes (Wahl et al., 1974; Milman et al., 1977).
2. Both are X linked (Westerveld et al., 1972; Seegmiller, 1976).
3. Hybrid combinations between Lesch–Nyhan cells and thioguanine-resistant permanent lines fail to complement the enzyme deficiency (Chapter 13).
4. Both are closely linked to the same genetic markers (Shows and Brown, 1975).

HPRT deficiency was one of the first characters for which the mutation rate in cell culture was measured (Szybalski and Szybalska, 1962). There are many estimates ranging over several orders of magnitude (Chapter 1). Little information is available on the properties of the first-step, low-level variants—in view of the fact that they are unstable and difficult to work with, this is not surprising. The tremendous variation in mutation rates for this character as measured in different systems requires some comment. There are several explanations for this wide range, some of them theoretical and some of them technical in nature. In the first place, it has only lately been appreciated that azaguanine is a less stringent selecting agent than thioguanine. This fact is due to the following: (1) azaguanine is a much poorer substrate for HPRT than is thioguanine (Van Diggelen et al., 1979); (2) guanases, active against azaguanine but not thioguanine, are present in serum (Ellis and Goldberg, 1972); and (3) in fetal calf serum there are substantial quantities of hypoxanthine, which will interfere with the efficiency of the selecting agents. Furthermore, the appreciation of metabolic cooperation in cultured cells has encouraged the use of appropriate controls and lower cell densities.

The phenomenon of metabolic cooperation has had a significant effect on the isolation of drug-resistant variants. It is a form of intercellular communication in which a particular mutant phenotype is corrected by the wild-type allele (Yotti et al., 1979). Examples of characters subject to metabolic cooperation include HPRT deficiency (Subak-Sharpe et al., 1969) and adenine phosphoribosyltransferase (APRT; EC 2.4.2.7) deficiency (Cox et al., 1972). Other characters, such as G6PD deficiency, are not corrected (Cox et al., 1972). The basis of metabolic cooperation resides in the transfer of nucleotides between cells through specific, permeable, intercellular gap junctions (Goldfarb et al., 1974). Metabolic cooperation has usually been detected as the phenotypic modification of mutant cells when in contact with wild-type cells (Fig. 2.4). Thus, when thioguanine-resistant cells are plated in the presence of wild-type cells, the transfer of the fraudulent nucleotide will substantially decrease the yield of resistant colonies. Investigations on a cell variant that does not participate in metabolic cooperation (mec⁻) have shown that such cells lack intercellular gap junctions, although treatment of the cells with dibutyryl cyclic adenosine monophosphate (db-cAMP) results in the restoration of both the junctions and the metabolic cooperation (Wright *et al.*, 1976).

To return to the question of the great variability in mutation rates to azaguanine or thioguanine resistance, one possible factor is the numbers of active Xs in different cell lines. Some workers have encountered great difficulty in selecting for azaguanine resistance in EUE, a HeLa derivative, whereas in mouse and hamster cells, HPRT⁻ lines can be obtained with ease. It was noted (L. L. DeCarli *et al.*, unpublished) that a correlation existed between the level of resistance to azaguanine and the number of chromosomes in the size range of the X; this finding suggests that resistance could be achieved in a stepwise fashion

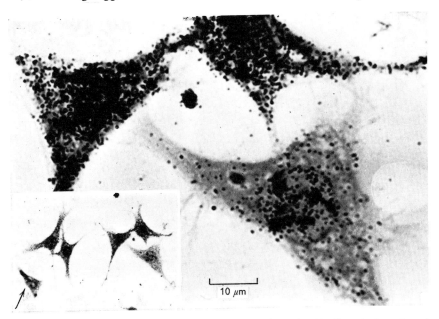

10 μm

Fig. 2.4. Metabolic cooperation in a coculture of transformed BHK cells and a line selected for its inability to carry out metabolic cooperation (mec⁻). Although the mec⁻ mutant was selected for its inability to transfer deoxypyrimidine nucleotides, it will still transfer purine bases, as shown by this study, using tritiated hypoxanthine. The fact that the two heavily labeled cells (upper left and center) are transferring material into the lightly labeled cell (lower right) established that the metabolic cooperation phenomenon shows a specificity for the substances transferred (Goldfarb *et al.*, 1974).

through loss of increasing numbers of X chromosomes. Raskind and Gartler (1978) have shown that in two permanent mouse fibroblast lines, one with one active X and one with two active Xs, induced mutation frequencies to thioguanine resistance were much higher in the former. This observation substantiates the belief that variation in the number of X chromosomes could profoundly affect the mutation rate.

Whatever the basis for the differences in mutation rate to HPRT deficiency in different cell lines, it does not appear to be a reflection of a greater genetic stability in normal diploid cell lines as compared to transformed malignant cell lines. For instance (see Chapter 1), the mutation rate to azaguanine resistance in diploid fibroblasts measured by DeMars and Held (1972) was 4.1×10^{-6}, actually higher than many of the rates measured in several transformed, permanent cell lines.

Perhaps one of the most useful aspects of the HPRT$^-$ variant is the fact that it can be selected in both directions. This was first pointed out years ago by Hakala (1957), was later exploited by Szybalski and his collaborators (1962), and has since been proved to be an extremely useful technique in a wide variety of experimental designs. This system takes advantage of the fact that HPRT$^-$ cells cannot utilize exogenous hypoxanthine and thus will perish when their *de novo* synthesis is extinguished with aminopterin or other inhibitors of one-carbon metabolism. In a medium containing hypoxanthine, aminopterin, and thymidine (HAT medium), only a cell that has recovered a functional enzyme will survive.

An interesting aspect of the HPRT$^-$ character has received substantial attention in recent years: the variation in mutation rates to azaguanine resistance at different ploidy levels (M. Harris, 1971). Harris's findings are surprising; because the trait is known to be recessive in both somatic cell hybrids and in human pedigrees, it would be expected that the mutation rate in tetraploids would be the square of that of the diploid and that the mutation rate in the octaploid would be the square of the tetraploid. No difference in mutation rate was encountered (see Table 2.2).

Harris's observations have been the subject of a good deal of speculation and have stimulated substantial research in several laboratories. Although at present there is no simple mechanism by which his data may be explained, several salient points merit consideration. In the case of thioguanine resistance, spontaneous mutation rates (Morrow *et al.*, 1978) and induced frequencies (Van Zeeland and Simons, 1975; Chasin, 1973) show the expected quantitative relationships in diploid and tetraploid hamster lines. Furthermore, Farrell and Worton (1977) have shown that in allotetraploid hamster clones possessing both a mutant and a wild-type HPRT allele, the occurrence of thioguanine-resistant clones could be accounted for by chromosomal segregation or, in some cases, by loss of a portion of a chromosome. Thus, it appears that the more stringent selecting agent, thioguanine, is able to select structural gene mutations in somatic cells and that a

TABLE 2.2   Mutation Rates to Azaguanine Resistance in Sublines of V79 Cells with Different Levels of Ploidy[a]

Cell line	Modal chromosome number	Assumed genome number	Mutation[b] rate $\times 10^5$
V5	21	1s	2.2
V25	43.5	2s	4.7
V68	83	4s	1.3

[a] Modified from M. Harris (1971).
[b] Mutations per cell per generation.

combination of mutation and chromosome segregation will account for the events observed in cell hybrids.

For azaguanine, the situation is less clear. Several authors have observed that in hamster or mouse cells a variety of resistant types occur. Sharp *et al.* (1973) and Morrow *et al.* (1973) found that azaguanine selects for a spectrum of types with HPRT activities varying from 0 to 100% that of wild type. Graf *et al.* (1976) has observed thioguanine-resistant clones of rat hepatoma cells that are purine overproducers and are suspected of possessing a mutation in the HPRT molecule with regulatory effects. Carson *et al.* (1974) demonstrated that a substantial fraction of clones selected in low concentrations of azaguanine were nonresistant phenocopies that escaped the selection procedure. Finally, in two instances (mouse and rat azaguanine resistance), it has been suggested that loss of the enzyme is due to a regulatory gene mutation (Watson *et al.*, 1972; Shin, 1974; Croce *et al.*, 1973a). This interpretation has, however, been questioned by Chasin and Urlaub (1976). Thus, it appears that although thioguanine selects for mainly HPRT$^-$ variants, azaguanine selects for a heterogeneous group, including the HPRT$^-$ class as well as other types whose biochemical basis may be complex. This would explain M. Harris's (1971) results since dominant mutations that may well occur in the case of azaguanine resistance would not be expected to vary in frequency as a function of ploidy.

At this time, the only mechanism of resistance to guanine and hypoxanthine analogs that is well defined is the loss or decrease in the level of HPRT. Although early experiments indicated that transport of purines in mammalian cells was linked to phosphorylation, more recent work has indicated that purines are transported into the cell by facilitated diffusion which precedes phosphorylation (Zylka and Plagemann, 1975). These latter studies utilized either HPRT$^-$ or energy-depleted Novikoff cells in order to avoid confusion of transport with phosphorylation of hypoxanthine by HPRT. Because purines are rapidly converted to their corresponding nucleotide and because such phosphorylated molecules are unable to move out of the cell, such activity can be erroneously equated with

SAINT PETER'S COLLEGE LIBRARY
JERSEY CITY, NEW JERSEY 07306

transport. It was found that cells lacking phosphoribosylation transport purines into the cell at the same rate as wild-type or nondepleted cells and that the equilibrium with the exterior is achieved in a matter of seconds.

The HPRT[+], azaguanine-resistant mutants described by Morrow et al. (1973; Table 2.3) have been studied in detail by Parsons et al. (1976). All show decreased incorporation of various purines into acid-soluble and acid-insoluble fractions, which suggests that these variants possess an altered enzyme that accepts the substrates with difficulty. This suggestion has been confirmed by Meyers et al. (1980), who have isolated large numbers of HAT-resistant, azaguanine-resistant mutants through a double selection procedure utilizing both substances. Such mutants are structural mutations that specifically alter the substrate recognition sites of the HPRT molecule.

In hamster cells, mutants with physically altered HPRT exist. These alterations include heat sensitivity and kinetic and electrophoretic properties (Fenwick, et al., 1977). Milman et al. (1977) have shown that a HeLa cell resistant to azaguanine possessed an altered peptide in its HPRT, thus providing definitive proof for the structural gene mutational hypothesis. Capecchi et al. (1977) have established that an HPRT mutant regains HPRT activity when ochre-suppressor tRNA is injected into it and that it is therefore a nonsense mutant.

Cell hybrids between mouse cells that are thioguanine resistant and lines that are azaguanine resistant will complement in hybrid combinations. Therefore, at least two cistrons exist, a fact that has also been established in hamster cells (Sekiguchi and Sekiguchi, 1973).

TABLE 2.3   Azaguanine (AG)-Resistant Clones Not Deficient in HPRT[a]

Cell line	Method of isolation	Cloning efficiency in HAT compared to control[b]	Level of azaguanine at which growth is inhibited 50%[c]	Specific activity in $m\mu M/mg$ protein HPRT
CHO/PRO[−]	None	1	3.3 ($\mu g/ml$)	1.19 (3)[d]
CHO 25	AG → HAT	1	7.5	1.60 (3)
CHO 27	AG → HAT	1	10.0	0.45 (3)
CHO 28	AG → HAT	1	60.0	3.3 (2)
CHO 22	AG → HAT	1	7.0	0.61 (1)
CHO 3	AG → HAT	1	Not determined	0.68 (1)
CHO 29	AG → HAT	1	32.0	1.35 (1)
CHO R1	AG	6×10^{-6}	21.0	<0.05 (1)
WAG[e]	AG	10^{-6}	Not determined	<0.05 (1)

[a] From Morrow et al. (1973).
[b] Cloning efficiency of 10^3 CHO/PRO[−] cells in HAT equals 45–55%.
[c] Fifty percent inhibition of growth interpolated from 4-day counts.
[d] Number in parentheses equals number of separate assays performed.
[e] Azaguanine-resistant derivatives of Walker rat carcinosarcoma.

Thus, elegant studies at the molecular level have established that at least some cases of azaguanine and thioguanine resistance result from mutational alterations within a structural gene. This simple statement does not, however, take into account the many unanswered aspects of the problem: the low-level, unstable variants; the possibility of gene inactivation (Chapter 1); the possible regulatory mutations; the variants in which resistance does not appear to be related to a loss or diminution of the enzyme. These questions will provide ample challenge for researchers in the coming years.

B. Diaminopurine

An enzyme that exists in the pathway of adenine salvage is the homolog of HPRT, namely, APRT. Mammalian cell mutants with reduced levels of the enzyme can be selected with the adenine analogs diaminopurine, fluoroadenine, and azaadenine. The gene for this enzyme is located on an autosome (see Chapter 6) and therefore would be expected to occur in two doses. Because resistance behaves as a genetic recessive in cell hybrids, it would appear that resistant mutants should be difficult, if not impossible, to obtain in cultured cells. This model has been investigated using CHO cells by Jones and Sargent (1974) and by Chasin (1974), who reached similar conclusions (i.e., that the induced mutation frequencies fit quite well the hypothesis of two active alleles with an intermediate heterozygous step; Fig. 2.5). This model has also been confirmed by Steglich and DeMars (1982) who observed mutation rates in human fibroblasts from individuals heterozygous for a genetic deficiency involving APRT. They found that the frequency of mutagen-induced DAP resistance in wild-type homozygotes was approximately the square of the induced frequency in heterozygous strains. Studies on APRT deficiency by Taylor *et al.* (1977) and Witney and Taylor (1978) using another CHO isolate have shown that (1) uptake appears to be separate from phosphorylation, (2) only one active allele appears to be present in wild-type cells, and (3) overproduction of purines does not occur in APRT⁻

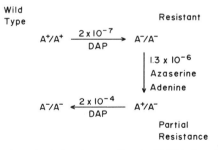

Fig. 2.5. Interpretation of ethylmethane sulfonate (EMS)-induced mutation frequencies. DAP, Diaminopurine. (From Chasin, 1974.)

cells in contradistinction to the situation in HPRT⁻ Lesch–Nyhan cells. Finally, Dickerman and Tischfield (1978) have published results of an interesting series of reconstruction experiments in which they utilized artificial mixtures of AP-RT⁻ and APRT⁺ cells. Their investigations show that for the three different analogs metabolized by APRT, much more inhibition of colony-forming ability in the presence of wild-type cells occurred in APRT⁻ variants in the presence of fluoroadenine than in the presence of diaminopurine. The reason for this difference is not clear, but it may be the result of the production of highly toxic intermediates from fluoroadenine by wild-type cells which diffuse into adjoining mutant cells.

Although the evidence accumulating indicates that APRT-deficient cell lines arise (in at least some cases) from mutations in the structural gene for the enzyme, there are instances of instability comparable to that of HPRT, and the issue of the origin of these variants is far from resolved. For instance, Bradley and Letovanec (1982) have characterized CHO hamster cells resistant to intermediate levels of diaminopurine, and have concluded that complete loss of the enzyme from the intermediate state can be achieved through two routes. One of these appears to be a standard mutational process, while the other occurs with a high rate and may reflect another molecular mechanism such as hypomethylation, whose nature is not yet completely understood. Gene transfer experiments utilizing the APRT locus (Wigler *et al.*, 1979) should soon clarify many of these problems.

III. PYRIMIDINE ANALOGS

The situation with regard to resistance to pyrimidine analogs (Figure 2.6) is comparable to that of the purines. In the best understood cases, resistance appears to be due to the loss or diminution of an appropriate nucleotide phosphokinase (most mammalian cells lack the ability to convert a pyrimidine base to a nucleoside). Thus, analogs of thymidine such as bromodeoxyuridine (BUdR) are converted by thymidine kinase (TK) to the mononucleotide form and thence into DNA, where they exert their toxic effect.

Among the earliest studies on pyrimidine analog resistance were those of Kit and his collaborators (1963). The BUdR-resistant mouse L cell line, LMTK⁻, was isolated through extended cultivation of the parent line in increasing levels of BUdR over a period of 100 weeks, resulting in a line that is completely lacking in thymidine kinase (EC 2.7.1.75). The LMTK⁻ line fails to grow in the reverse selective HAT medium, and revertants to wild-type have never been reported. The extreme stability of the LMTK⁻ phenotype suggests that it is the result of a deletion and makes it quite useful for selection of cell hybrids and other experiments in genetic transfer (see subsequent chapters). Further studies

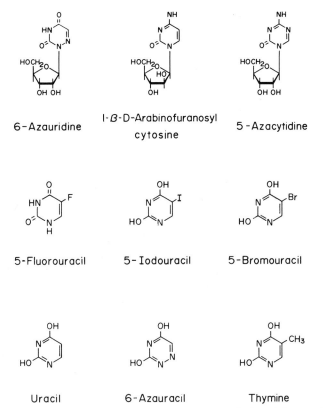

6-Azauridine   1-β-D-Arabinofuranosyl cytosine   5-Azacytidine

5-Fluorouracil   5-Iodouracil   5-Bromouracil

Uracil   6-Azauracil   Thymine

Fig. 2.6. Structures of some naturally occurring pyrimidines and their analogs.

by Kit and Leung (1974) have shown that another thymidine kinase exists that is specific to the mitochondria. In normal cells it represents approximately 10% of the total activity, and it can be distinguished by disc polyacrylamide gel electrophoresis (PAGE) analysis. It is not affected in BUdR-resistant lines and appears to be controlled by a gene on a different chromosome.

The gene for thymidine kinase has been localized to the human E17 chromosome (see subsequent chapters) and therefore would be expected to appear in two doses, all other things being equal; this double dosage possibly explains the difficulty encountered in selecting BUdR-resistant lines (Littlefield, 1965). This question has been considered quantitatively by Clive *et al.* (1972) in mouse lymphoma cells and by Roufa *et al.* (1973) in V79 Chinese hamster cells. Their studies have shown that even when as many as 10^9 cells are examined, TK$^-$ mutants cannot be obtained directly. However, through stepwise selection it is possible to isolate TK$^-$ lines; this finding suggests the existence of an intermedi-

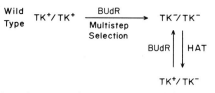

Fig. 2.7. Interpretation of mutation frequencies to BUdR resistance. (From Roufa *et al.,* 1973.)

ate, heterozygous state (Fig. 2.7). Through the use of HAT medium, a number of revertants have been obtained and characterized (Table 2.4), and they appear to have only partial recovery of activity, a result that is also in agreement with the heterozygosity model. This interpretation has, however, been questioned by Harris and Collier (1980). These authors have suggested that BUdR resistance in hamster cells comes about through very small increments, rather than through single step jumps in the level of resistance, as required by the heterozygosity model. Their data were obtained by growing populations of cells in low levels of the drug and testing them at various intervals for TK levels, HAT resistance, and BUdR resistance. Their observations do not appear to be reconcilable with a simple model involving two alleles at a single locus.

Freed and Mezger-Freed (1973) have obtained results in haploid frog cells that

TABLE 2.4   Thymidine Kinase and Karyotype Properties of Chinese Hamster Clones[a]

Clone	Karyotype (number of chromosomes)	Thymidine kinase-specific activity
A. Wild type V79	21	3.30
B. TK⁻ clone 462-10	18	0
C. TK⁻ revertants[b]		
D1	36	0.362
D2	19	0.560
D4	19	0.092
D6	18	0.230
D7	19	0.078
D8	19	0.092
D9	19	0.089
D10	19	0.076

[a] From Roufa *et al.* (1973).
[b] Revertant TK⁺ clones selected from 462-10 (TK⁻) in HAT medium are termed D1–D10. These revertants have been analyzed for karyotype number and thymidine kinase. Thymidine kinase-specific activities are expressed as picomoles [^{14}C]TMP per minute per microgram of extract protein.

also are not immediately reconcilable with the heterozygosity model. They have detected an intermediate step in the acquisition of BUdR resistance; however, they ascribe this to the loss of a specific thymidine transport system, which appears to be a requisite intermediate in the development of the TK⁻ phenotype. Mutants resistant to high levels of thymidine (which apparently is toxic due to feedback inhibition to cytidine synthesis) can also be obtained in mammalian cells (Breslow and Goldsby, 1969). Such mutants occur with a mutation rate of 2.6×10^{-4} as measured by the fluctuation test, and their uptake of thymidine is depressed, suggesting an effect on transport as the basis of resistance.

Freed and Hames (1976) have suggested that there may exist (at least in frog cells) two forms of TK: a thermolabile form that is predominant and whose loss is required for acquisition of BUdR resistance, and a thermostable type, which is also the transport protein. This model would be compatible with the results of Breslow and Goldsby (1969), who did indeed find a slight depression in TK activity in their thymidine-resistant mutants which presumably represented the loss of transport kinase. However, it may be that these authors and several other workers have confused transport with uptake, and that they have measured the ability of the cells to trap radioactive molecules rather than to transport them across the cell membrane. Wohlhueter and Plagemann (1980) designate transport as the mediated permeation of a substrate molecule by a specific saturable mechanism. Furthermore, they define uptake simply as radioactivity appearing within the cell which has its origin in some exogenous substrate. Since the transport of purine and pyrimidine bases and nucleosides into mammalian cells is a relatively nonspecific process, the rate-limiting step in uptake will be phosphorylation rather than transport of these molecules. Equilibrium between the outside and the inside of the cell may be achieved within a matter of seconds, and the character-

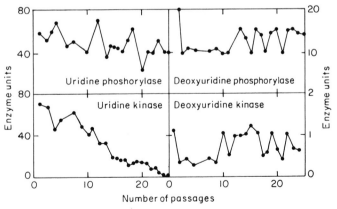

Fig. 2.8. Activity of enzymes involved in the metabolism of uridine during treatment of mice carrying Ehrlich ascites tumors with 5-fluorouracil (Veselý and Čihák, 1973).

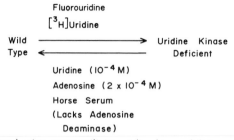

Fig. 2.9. Reverse selective system utilizing uridine kinase deficiency. (From Medrano and Green, 1974.)

ization of this process requires a rapid sampling technique (Wohlhueter *et al.*, 1979). When such techniques are utilized, it has been found that (1) transport and phosphorylation are clearly separable processes with very different rates and specificities, and (2) none of the purine or pyrimidine analog-resistant mutants involves altered transport. At this time, the burden of proof falls on one who would claim the existence of such a variant.

Another pyrimidine analog-resistance system showing some relationship to thymidine resistance is arabinofuranosyl cytosine resistance. When converted to the triphosphate form, this compound is a potent inhibitor of DNA synthesis. Resistance to this antileukemic nucleoside is associated with at least two different mechanisms: (1) loss of deoxycytidine kinase, which behaves as a recessive in cell hybrids (de Saint Vincent and Buttin, 1979); (2) expansion of the dCTP pool, which dilutes out the effect of the drug and results in cross resistance to thymidine. In both cases, the genetic alterations appear to be accountable on the basis of classic mutations (i.e., base pair changes in nuclear genes).

A number of toxic pyrimidine analogs can be used to select variants with depressed levels of uridine kinase (Veselý and Čihák, 1973). These include azauridine, fluorouracil, azacytidine, and fluorouridine. Typical is the response shown in mice carrying Ehrlich ascites tumor cells during treatment with 5-fluorouracil (Fig. 2.8).

Medrano and Green (1974) have developed a reverse selective system for uridine kinase-deficient mutants similar to the HAT selective system (Fig. 2.9).

IV. DRUGS OTHER THAN PURINE AND PYRIMIDINE ANALOGS

A. Ouabain

This steroid-like compound is an inhibitor of a Na^+,K^+-ATPase (EC 3.6.1.3) located on the cell membrane. It is ultimately cytotoxic, and mutants resistant to its effects can be isolated in a single step. Baker *et al.* (1974) isolated both mouse and Chinese hamster cells resistant to ouabain with a spontaneous rate of 5×10^{-8} per cell per generation. Such mutants are stable in the absence of the selecting agent and behave as dominant mutations in hybrid combinations. Diploid human fibroblasts resistant to ouabain have also been isolated and their resistance appears to be due to an altered response of their Na^+,K^+-ATPase to inhibition by ouabain. The most parsimonious hypothesis for the genetic basis of this alteration is a mutation in the DNA of the structural gene specifying the Na^+,K^+-ATPase. These variants lend themselves quite successfully to a selective system combined with HPRT deficiency. A cell doubly resistant to thioguanine and ouabain can be hybridized with a cell lacking any selective markers, because the $HPRT^-$ is a recessive and the Oua^r is a dominant. Thus, in the presence of HAT plus ouabain, both parent cells will die and only hybrids will emerge.

B. Plant Lectins

These substances possess numerous biological effects resulting from specific interactions with carbohydrate moieties of cell surface glycoproteins. They are often cytotoxic and can be used in the study of drug resistance. Phytohemagglutinin (PHA), concanavalin A (Con A), wheat germ, ricin, and *Lens culinarius* agglutinin are some of the commonly employed lectins. Resistant mutants that have been isolated possess cell surface alterations due to absence or reduction of glycosyltransferase (Table 2.5). Phytohemagglutinin-resistant lines are cross-sensitized to Con A and cross-resistant to other lectins (Stanley *et al.,* 1975a). The resistant phenotype is recessive in cell hybrids, occurs with a mutation rate of 1.5×10^{-6} (Chapter 1), and its resistance behaves as a single-step mutation (see Fig. 2.10). Mutagens are effective in increasing the frequency of resistant clones over the spontaneous mutation rate. Furthermore, the results of the fluctuation test prove that the mutants are preexisting in the population and have lost the enzyme UDP-*N*-acetylglucosamine–glycoprotein *N*-acetylglucosaminyltransferase (Stanley *et al.,* 1975b).

Krag (1979) and Cifone *et al.* (1979) have isolated Con A-resistant mutants in CHO cells that represent a different class of variants from those considered by Stanley *et al.* (1975a,b). The mutations behave as recessives in cell hybrids and show substantially reduced incorporation of mannose into glycoproteins and into lipid-linked oligosaccharides. Con A-resistant mutants are unable to glucosylate oligosaccharide–lipid at the same rate as wild-type cells, as shown by experiments measuring the transfer of [^{3}H]glucose to oligosaccharide–lipid. Thus, the basis for the resistant phenotype appears to be an incomplete assembly of membrane glycoproteins.

TABLE 2.5 Glycosyltransferase Activity of Wild-Type (WT) and Phytohemagglutinin-Resistant (Pha^R) CHO Cells[a]

Glycosyltransferase substrates		Number of experiments	Mean glycosyltransferase activities[b] (nmol/mg of protein per hr)		% WT
Nucleotide-sugar	Exogenous acceptor		WT	Pha^R	
CMP-sialic acid	AGP(-SA)	8	5.4 ± 0.90^c	6.1 ± 0.91^c	—
GDPfucose	AGP(-SA, Gal)	7	11.6 ± 3.3^c	9.5 ± 1.1^c	—
UDPgalactose	GlcNAc	6	18.7 ± 2.6^c	19.5 ± 3.1^c	—
UDPgalactose	AGP(-SA, Gal)	8	9.2 ± 1.3^c	9.5 ± 1.5^c	—
GDPmannose	None	3	0.28 ± 0.06^c	0.27 ± 0.07^c	—
UDPGlcNAc	AGP(-SA, Gal, GlcNAc)	15	12.3 ± 3.0^d	6.72 ± 1.8^d	55[d]
UDPGlcNAc	RNase B	7	3.7 ± 0.91^d	0.94 ± 0.39^d	25[d]
UDPGlcNAc	IgG(-SA, Gal, GlcNAc) glycopeptide	2	4.8	0.2	4
UDPGlcNAc	AGP(-SA, Gal, GlcNAc) glycopeptide	2	7.8	3.0	39

[a] From Stanley et al. (1975b).
[b] The arithmetic means ± standard deviations of the enzyme activities from several separate experiments are presented.
[c] Not significant at $p < 0.05$.
[d] Significant at $p < 0.001$.

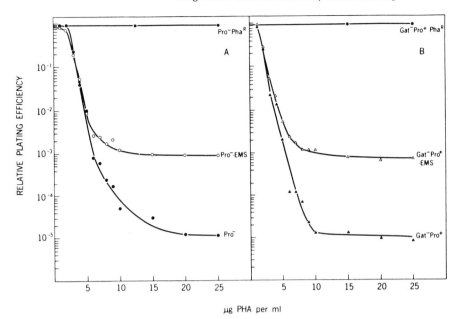

Fig. 2.10. Survival of mutagenized, unmutagenized, and resistant CHO cells plated in the presence of phytohemagglutinin. Suspensions of Pro⁻ cells (A) and Gat⁻ Pro⁺ cells (B) were divided and half were treated with EMS. After 18 hours the cells were washed, resuspended in fresh medium, and returned to suspension culture. After a further 3 days, the mutagenized and unmutagenized cells were plated in various concentrations of PHA and incubated for 8 days (A) or 9 days (B) at 37°C. The relative plating efficiencies of typical colonies that were selected in PHA are also shown (Stanley et al., 1975a).

C. Folic Acid Analogs (Amethopterin and Aminopterin)

These substances are inhibitors of the enzyme folic acid reductase (see next chapter) and kill the cell by inhibiting one-carbon metabolism and the *de novo* pathway of purine and pyrimidine synthesis. Dihydrofolate (DHF, the substrate of the reaction) is reduced to tetrahydrofolate to supply one-carbon units for purine and thymidylate synthesis. Figure 3.5 (see Chapter 3) shows the structure of folic acid and some of its antagonists. They have proved to be quite useful in cancer chemotherapy, and for this reason, as well as their utility in enlarging our theoretical knowledge, an understanding of resistance to these substances is of great interest. Resistance can be due to several mechanisms: (1) a mutation affecting permeability, (2) a mutation causing an alteration in the enzyme dihydrofolate reductase, or (3) an increase in the level of the enzyme. The third mechanism of resistance is the most dramatic and the best understood on the molecular level. In such lines the enzyme activities are elevated as much as

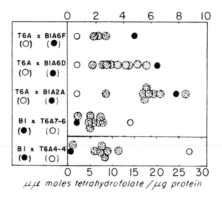

Fig. 2.11. Folic acid reductase activities of wild-type, sensitive parent (open circles); amethopterin-resistant parents (closed circles); and hybrids (dotted circles). (From Littlefield, 1969.)

3000-fold and result from an apparent increase in the number of molecules (a conclusion based on titration with aminopterin, purification of the enzyme, and electrophoretic and kinetic studies).

Hybridization studies (Littlefield, 1969) have shown that hybrids between wild-type and overproducing lines are intermediate in activity; therefore, the increase in activity is not due to the loss of a diffusible repressor (Fig. 2.11).

Alt *et al.* (1978) have demonstrated that in at least two cases overproduction of dihydrofolate reductase in response to methotrexate challenge is a result of selective multiplication of the structural gene for the enzyme (Table 2.6). In these elegant studies the authors first obtained specific mRNA for the enzyme by immunoprecipitation of dehydrofolate reductase-synthesizing polysomes. A

TABLE 2.6   Relative Level of Dihydrofolate Reductase Activity, mRNA, and Gene Copies in S-180 and L1210 Lines[a]

	Relative dihydrofolate reductase		
Line	Specific activity[b]	mRNA sequences	Gene copies
S-180			
S-3	1	1	1
AT-3000	250	220	180
Rev-400	10	7	10
L1210			
S	1		1
RR(+mtx)	35		45
RR(−mtx)	35		35

[a] From Alt *et al.* (1978).

[b] In each column, values are normalized to those of the sensitive line that was taken as 1.

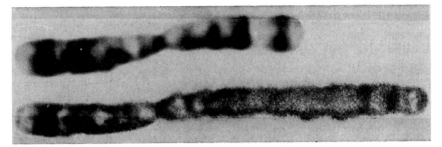

Fig. 2.12. Expanded chromosome associated with methotrexate resistance. Chromosome 2 from a drug-sensitive (top) and a stably drug-resistant cell (bottom) are shown. Both are stained with Giemsa following procedures to demonstrate banding profiles. The homogeneously staining region harbors the redundant area of the chromosome in which the gene appears in approximately 200 copies. (From Schimke, 1980.)

complementary cDNA was prepared with reverse transcriptase and subsequently purified. Using the cDNA as a probe, association kinetics were carried out with DNA from wild-type, highly resistant, and revertant cell DNA. The results are consistent, demonstrating a one-to-one correlation between the ratio of the enzyme molecules, number of messenger molecules, and number of gene copies in wild-type and overproducing lines (Table 2.6). Recent efforts have been made to determine the chromosomal location of these allelic copies. The authors speculate that the long, non-Giemsa-staining chromosomal segments observed in hamster cells that are dihydrofolate overproducers may be a cytological manifestation of genetic redundancy (Biedler and Spengler, 1976; see Fig. 2.12). Further studies (Nunberg *et al.*, 1978) using *in situ* hybridization of labeled DNA have confirmed that the DHF reductase genes are specifically localized to the homogeneously staining regions.

D. Glucocorticoids

The molecular mechanism of toxicity of these substances is not known at present. They are potent enzyme inducers (see subsequent chapters); in hepatoma cells, dexamethasone (a synthetic glucocorticosteroid) penetrates the cell membrane and binds reversibly to a cytoplasmic receptor. It is then transported to the nucleus as a complex where it binds to the DNA as an allosterically modified hormone receptor complex and stimulates the accumulation of specific messenger RNAs (Rousseau, 1975).

Mouse lymphoma cells resistant to glucocorticoids have been isolated (Sibley and Tomkins, 1974). The mutation rate (Chapter 1) as measured by the fluctuation test was 3.5×10^{-6} per cell per generation, a figure that shows a substantial increase with mutagens. There is a gradual increase in the frequency of mutants over a period of several years. Because steroid resistance is recessive in somatic cell hybrids, the mutation rate must be quite high unless that part of the genome in question is functionally haploid.

Pfahl and Bourgeois (1980) have studied a number of dexamethasone-resistant mouse lymphoma cell lines that owe their phenotype to either an absence of receptors (r^-) or to reduced transfer into the nucleus (nt^-). Variants of the S49 lymphoma and thymoma cell lines with different mutations for dexamethasone resistance were isolated. In addition, these cell lines carried mutations to facilitate the isolation of hybrids (TK^- and $HPRT^-$). An analysis of such hybrids indicated that both r^- and nt^- are recessive to r^+ and that r^- and nt^- fail to complement and are therefore different mutations in the same gene.

E. Puromycin, Actinomycin D, and Colchicine Resistance

An interesting situation involving a pleiotropic variant is observed in the case of a mutation causing simultaneous resistance to a number of substances. In contradistinction to many of the other variants discussed here, the biochemical basis appears to reside in a generalized alteration of permeability of the cell membrane, encompassing a wide range of compounds (Table 2.7). It is noteworthy that a positive correlation exists between the partitioning of the drugs and resistance to them (Bech-Hansen et al., 1976). This observation has led these authors to suggest that resistance may be due to an alteration of membrane fluidity, the molecular basis of which resides in the appearance of a new glycoprotein in the membrane of resistant cells (Ling, 1975), shown by studies using [^{3}H]borohydride or [^{3}H]glucosamine. Studies with cyanide inhibition indicate that this barrier is energy dependent (See et al., 1974) and, as might be expected for such an alteration, appears to be codominantly expressed in cell hybrids (Ling and Baker, 1978).

The genetic basis of this pleiotropic character is somewhat obscure at present. Puromycin resistance was one of the first phenotypes for which mutation rates were measured (Lieberman and Ove, 1959) using the fluctuation test (see Table 1.3). Resistance appears to be stable in some situations (Ling, 1975); however, in other cases, resistant lines revert rapidly when removed from the selective agent (Morrow et al., 1980). The stepwise pattern of resistance encountered suggests a system of multiple genes or alleles. This observation combined with instability and reversion is quite reminiscent of antifolate resistance. It is quite easy to conceive of a protein present on the cell membrane in small amounts that might be augmented through gene duplication and consequent increase in the quantity of messenger and gene product. Although appealing, this hypothesis does not appear to be testable at present because we have inadequate information on the nature of the primary gene product.

Ling et al. (1979) have isolated a different class of colchicine-resistant mu-

TABLE 2.7 Cross-Resistance of a Highly Colchicine-Resistant
Derivative of the CHO Cell Line to a Variety of
Compounds[a]

Drug	Relative cross-resistance[b]
Colchicine	184
Puromycin	105
Daunomycin	76
Emetine	29
Ethidium bromide	11
Acriflavin	7
TBA	7
TPMP	9
Cytochalasin B	11
Erythromycin	5
Colcemid	16
Vinblastine	29
Gramicidin D	144
Adriamycin	25
Proflavin	4
Procaine	0.4[c]
Tetracaine	0.2
Xylocaine	0.1
Propanolol	0.2
5β Pregnan-3,20-dione	0.1
Deoxycorticosterone	0.1
1-Dehydrotestosterone	0.1

[a] Data are composites from Bech-Hansen et al. (1976).

[b] Values represent the dose of the drug that inhibits growth by 50% in the resistant cell, corrected to a value of unity for the parent line.

[c] In some cases the value is less than unity, a situation referred to as collateral sensitivity.

tants through an approach that circumvents the permeability barrier. Using Tween 80 to potentiate drug entry into the cells, they have obtained a class of colcemid-resistant mutants that are not affected in their transport of this or related compounds but rather produce a form of tubulin with reduced colchicine-binding ability. These mutants may be quite useful in analyzing the structure of microtubules.

F. Emetine Resistance

Emetine is another inhibitor of protein synthesis that has been successfully employed in the isolation of drug-resistant mammalian cells. Chinese hamster ovary cells showing 10- to 50-fold increases in resistance to the drug have been characterized, and the basis of resistance has been localized to an alteration in the 40 S ribosomal subunit. From these variants, a higher, second level of resistance has been obtained; the phenotype of these second-order variants also derives from a lesion in the polyribosomal fraction. Both mutants are discrete steps and both recessive. Emetine-resistant mutants are cross-resistant to other members of the ipecac family of alkaloid protein synthesis inhibitors. Thus, these substances may prove to be of value in the genetic analysis of the ribosomal structure (Gupta and Siminovitch, 1978b).

This possibility has been substantiated by Reichenbecher and Caskey (1979) and Boersma *et al.* (1979), who have carried out biochemical characterizations of emetine-resistant cell lines. Both groups have resolved the ribosomal proteins of CHO cells through two-dimensional polyacrylamide gel electrophoresis and have determined that in emetine-resistant cell lines a protein known as s22 displays an altered mobility. This protein possesses a different charge but not a different molecular weight, suggesting that the change is the result of a point mutation.

Great differences exist in mutation rates for emetine resistance between cell lines. This observation has led to the suggestion that the CHO line, which shows a high mutation rate to resistance, is functionally hemizygous for the emetine locus. If this were true, then one would expect that the rate of segregation of emetine resistance in a CHO[r] × CHO[s] hybrid would be much higher than in a CHO[r] × presumptive functional diploid[s]; and, in fact, this is what is observed (Gupta *et al.*, 1978b). However, there are some difficulties with the value of the quantitative estimates that lead to the suspicion that other factors may be operating in the generation of variability for these characters (see Section V).

G. α-Amanitin

The bicyclic octapeptide α-amanitin (derived from the mushroom *Amanita phalloides*), when used at low concentrations, is a specific inhibitor of DNA-dependent RNA polymerase II (EC 2.7.7.6), the enzyme responsible for chromatin transcription. Resistant hamster cells whose phenotype is due to an alteration in the structure of the enzyme, rendering its refractory to inhibition, have been isolated (Chan *et al.*, 1972). On the other hand, Somers *et al.* (1975) have isolated an α-amanitin-resistant line that is apparently an overproducer of the enzyme. Gupta and Taylor (1978) have reported a partially resistant CHO variant that possesses a modified RNA polymerase. In addition, these mutants appear to have alterations in the cell membrane that render them more sensitive to colchicine and cordycepin inhibition.

α-Amanitin resistance possesses a number of properties that suggest that it results from a structural gene mutation: low spontaneous mutation rate, alterations in the physicochemical properties of the responsible enzyme, and stability in the absence of the selecting agent. The resistance acts as a codominant in somatic cell hybrids because both enzyme types are expressed. All of this is in agreement with a simple mutational model involving base pair changes in the structural gene of the responsible enzyme (the mutants of Somers *et al.*, 1975, are more problematical in this regard).

The ease with which α-amanitin-resistant mutants have been isolated in CHO cells (Gupta and Siminovitch, 1978a) and in L6 myoblasts (including temperature-sensitive mutants; Ingles, 1978) also indicates that the gene involved is present in only one functional dose (Ingles *et al.*, 1976). This high mutation rate is especially striking in view of the fact that the temperature-sensitive mutant phenotypes are recessive in hybrids. Because the CHO cell has most of its diploid complement of DNA, it would appear that substantial segments of the genome may be inactivated.

H. Chloramphenicol Resistance

Chloramphenicol (CAP) is a mitochondrial protein synthesis inhibitor that can be used in the isolation of drug-resistant lines (Mitchell *et al.*, 1975). Chloramphenicol resistance has been used as a marker in experiments in which hybrid combinations of nuclei and cytoplasms were obtained with the aid of cytochalasin B and reassembled using polyethylene glycol (PEG)-mediated fusion (Wallace *et al.*, 1975). These experiments will be discussed in detail in subsequent chapters; however, it should be stated that chloramphenicol resistance is a cytoplasmically determined character.

I. Diphtheria Toxin

Diphtheria toxin, a protein of molecular weight 63,000, is an extremely specific and potent inhibitor of protein synthesis in eukaryotic cells and is composed of two subunits (Collier, 1975). One of these (the B chain) interacts with receptors on the cell surface. The other (A chain) reacts with the protein synthesis elongation factor 2 (EF-2) by attaching an ADPribose to it; this modification prevents its functioning. The attachment occurs at a single amino acid, a modified histidine.

Several mechanisms of resistance have been described. One is a first-step level involving a decrease in the rate of transport of toxin into the cell (Gupta and Siminovitch, 1980). A second is an alteration in the EF-2 itself, presumably a mutation altering the amino acid inserted at the ribosylation-sensitive site (Gupta and Siminovitch, 1980). Finally, a third type of mutation would be one affecting the enzyme that modifies the histidine. Moehring *et al.* (1980) have shown that extracts from normal cells will convert this EF-2 molecule to a toxin-sensitive state, presumably by reacting the histidine to its modified form. The permeability mutation is recessive in hybrids, as is the mutation that affects the posttranslational histidine modification system. The EF-2 mutation is codominant.

V. CONCLUSIONS

Although at present a great deal of information is available on the specific biochemistry of drug resistance, the genetic basis of many variants is still unknown. Certainly, in the case of methotrexate resistance, the basis lies in an overproduction of the dihydrofolate reductase enzyme as a result of a genetic redundancy; furthermore, the HPRT-deficient variants studied by Capecchi *et al.* (1977) and Milman *et al.* (1977) are no doubt base pair substitutions in the structural gene for the enzyme. We still, however, lack information on the overall basis of variability of somatic cells. The fact that recessive mutations can be obtained with ease in a variety of cell lines has suggested to a number of authors that the genome of many mammalian cell lines may be functionally haploid. One test of this hypothesis was performed by Gupta *et al.* (1978b), in which emetine-resistant cells were hybridized to (1) cells that were believed to be haploid for a functional emetine sensitivity gene, and (2) to three other cell lines that were believed to be diploid for this same marker. The hybrids were plated in emetine, and the segregation rate was calculated using the Luria-Delbrück fluctuation test. The results (Table 2.8), although in the right direction, do not show good quantitative agreement with the hypothesis (see Chapter 1). For instance, the three hybrid combinations that were suspected of carrying two alleles for sensitivity should show segregation rates that are the square of the rate for the suspected hemizygote (the first cross). These data disagree with the prediction by three orders of magnitude, suggesting that if the model of functional hemi-

TABLE 2.8   Segregation Rates for Emt[r] in Various Hybrid Cell Lines[a]

Hybrid cell lines[b]	Presumed[c] emetine genotype	Segregation rate (number of segregants per cell per division)[d]
EOT-3xGat	r⁻R⁺	2.3×10^{-4}
EOT-3xM3-1	r⁻R⁺R⁺	3.2×10^{-5}
EOT-3xGm7s	r⁻R⁺R⁺	1.6×10^{-5}
EOT-3xV79	r⁻R⁺R⁺	3.2×10^{-5}

[a] Modified from Gupta *et al.* (1978b).

[b] The first hybrid is a CHO × CHO cross. The others are CHO crossed to hamster lines presumed to be diploid for the emetine locus.

[c] These designations are based on the fact that the Gat line has a much higher rate to emetine resistance than the other hamster cell lines.

[d] Obtained through a Luria-Delbrück fluctuation analysis.

zygosity for the original parent lines is correct, then segregation of one of the three functional chromosomes from any particular cell would increase the probability of a second segregational event. Until evidence in support of this contention can be mustered, the hemizygosity model remains speculative.

α-Amanitin resistance, in which some resistant lines have been shown to produce two forms of the enzyme, has been interpreted in terms of a heterozygosity model. In this hypothesis, the resistant lines carry one normal allele and one mutant allele, analogous to the proposal of Clive *et al.* (1972) to account for the pattern seen in the case of thymidine kinase variation or that of Chasin (1974) for DAP resistance. Our recent realization of two aspects of eukaryotic genetic structure suggests that there may be other interpretations. In the first place, the studies on the globin genes in mammals have shown that the α-globin gene occurs in two similar tandem copies (see Chapter 8). Thus, even in normal intact organisms, there may be a fair degree of genetic redundancy. Second, the situation with dihydrofolate reductase overproduction suggests that mammalian cell lines in culture may achieve a high degree of reiterative information with relative ease. Similarly, the high mutation rates discussed in Chapter 1 may reflect an overall degree of hypervariability in mammalian cells that results in a continuing loss, gain, and transposition of elements throughout the karyotype. Thus, the results in the case of α-amanitin resistance or BUdR resistance may reflect the presence of a second gene produced through duplication as a result of the sojourn *in vitro,* or a second functional gene that was present in the original organism in a manner similar to the hemoglobin situation (see Chapter 8).

Finally, it should be borne in mind that the possibilities discussed by Schimke (1980) may represent a general situation in mammalian cell variants. It is quite likely that a number of different genes conferring resistance to toxic agents are initially present on microchromosomes, which are highly unstable, and later through selective pressure become joined to a larger chromosome, the final union resulting in a phenotype stabilization.

Auxotrophic Variants in Cultured Cells

I. INTRODUCTION

Variation in the nutritional requirements of cultured cells was exploited at a later time than was drug resistance. This appears to be due to the technical problems encountered in the selection of nutritionally altered mammalian cells, including the complexity of tissue culture media and the presence of undefined molecules in the calf sera. In addition, mammalian cells already require most of the amino acids or are able to catalyze only a final reaction in their synthesis. Despite these disadvantages, however, the last 10 years have seen the characterization of a multitude of variants altered in the ability to synthesize low-molecular-weight metabolites in quantities sufficient for growth. The great bulk of evidence describing their properties favors the hypothesis that they are the result of mutations in structural genes for enzymes catalyzing the synthesis of metabolic intermediates. The evidence includes

1. The occurrence of mutations at low frequency with a stable phenotype in the absence of selection.
2. An increase in mutation frequency with exposure to mutagens.
3. Production of an altered gene product.
4. Localization to specific chromosomes.

In this chapter we shall consider the nature of this evidence and delve into the properties of specific types of auxotrophic variants.

II. TECHNIQUES FOR ISOLATION

It is not often practical to isolate rare, nutritionally deficient cells from a vast population of nonrequirers through direct means. Therefore, a method of enriching the mutant population must be utilized, and such an approach must operate to eliminate the wild-type cells while leaving the mutants intact. A model for such a technique is the penicillinase technique of Lederberg and Lederberg (1952), in which a bacterial population was plated in medium containing penicillin (which kills dividing cells) but lacking a particular metabolite. Later, the metabolite and the enzyme penicillinase were added simultaneously to the medium. Cells requiring the particular metabolite were saved from penicillin death because of their inability to divide in the absence of the metabolite. Subsequently, they proliferated in the presence of the particular nutrient for which they were auxotrophic and in the absence of penicillin, which was eliminated by the penicillinase. As might be imagined, this method is not totally efficient because a proportion of prototrophs escape the selection procedure, while at the same time a certain fraction of the desired auxotrophs are killed in the course of the selection. It would, in fact, be presumed that some classes of auxotrophs could never be isolated, namely, those that die in the absence of their requisite nutrient. For example, bacteria that require thymine undergo "thymineless death" in its absence, and one would therefore expect that thymidine-requiring strains could not be isolated in mammalian cells. However, this situation could be circumvented with an enrichment procedure that could kill off the prototrophs before the auxotrophs starved to death in the absence of their requirement. By serial repetition of selection, one might enrich over several cycles and gradually obtain a population with a high proportion of the desired auxotroph. As we will see, this approach does appear to be operable under certain conditions.

A reverse selection procedure for isolating auxotrophs in mammalian cells was first described by DeMars and Hooper (1960) and utilized aminopterin to eliminate growing cells followed by subsequent addition of the metabolite in question. A later refinement of this technique was described by Kao and Puck (1968; Fig. 3.1), one which takes advantage of the ability of dividing cells to incorporate BUdR and their sensitization to visible light. A variation of this approach is to use long-wavelength uv (Chu et al., 1972), which appears to give a better yield of mutants possibly because of a higher efficiency on the part of the procedure or, alternatively, because of a mutagenic action on the part of the long-wavelength uv itself. Representative data of Kao and Puck (1969) are reproduced in Table 3.1. It is necessary to test fairly substantial numbers of clones in order to obtain auxotrophic variants.

As an alternative to enrichment procedures for the isolation of mutants, a number of replicate plating methods have been developed. Such methods seek to semiautomate the transfer of large numbers of clones in order to sidestep the many laborious operations involved in isolating and testing thousands of individual colonies. Two general approaches to this problem have been utilized. The first uses devices that are essentially multipipetters and, through the use of microtiter plates and multiple syringes, can transfer as many as 100 colonies to fresh microtiter plates in a single series of operations (Goldsby and Zipser, 1969;

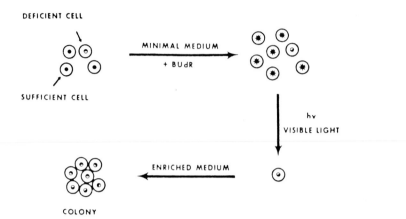

Fig. 3.1. Schematic representation of the BUdR–visible light technique for isolation of nutritionally deficient mutant clones. The mixed cell population is exposed to BUdR in a deficient medium in which only the prototrophs can grow. These alone incorporate BUdR into their DNA and are killed on subsequent exposure to a standard fluorescent lamp. The medium is then changed to a composition lacking BUdR and enriched with various nutrilites. The deficient cells grow up into colonies. (From Kao and Puck, 1968.)

Wild and Hellkuhl, 1976; Brenner *et al.*, 1975; Suzuki *et al.*, 1971). Alternatively, replication methods copied from Lederberg and Lederberg's (1952) original velvet method have been employed. Most of these techniques use filter paper (Esko and Raetz, 1978) or nylon (Stamato and Hohmann, 1975) disks, which are allowed to be in contact with growing colonies long enough for fragments of these colonies to detach so as to allow transfer to a new petri dish in the same pattern as the original. Although these techniques have enabled investigators to isolate rare mutants without enrichment, none of these procedures is as

TABLE 3.1   Yield of Deficient Mutants Obtained in Separate Experiments with Two Mutagens[a]

Mutagen	Number of clones surviving treatment with BUdR and visible light	Number tested	Mutants obtained
None	254	198	0
MNNG[b]	332	184	18
EMS[c]	373	175	29

[a] Data from Kao and Puck (1968).
[b] Survival with *N*-methyl-*N*-nitrosoguanidine (MNNG) = 26%.
[c] Survival with ethylmethane sulfonate = 78%.

simple or as foolproof as the technique of transferring yeast and bacterial colonies with the aid of sterile velvet.

III. PROPERTIES OF VARIANTS

A. Auxotrophs of Purine Metabolism

The *de novo* pathway of adenine synthesis (Fig. 3.2) contains a series of reactions, any one of which can be interrupted by mutation in a fashion similar to

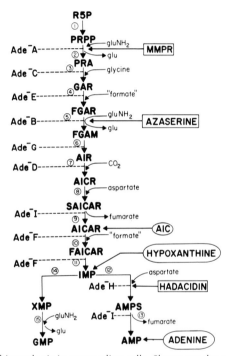

Fig. 3.2. Purine biosynthesis in mammalian cells. Shown are the reactions leading to the synthesis of adenylic acid. The steps are numbered 1 through 15, and the letters under the enzymes indicate the complementation group in CHO cells for that enzyme. Mutants for enzymes 8 and 11 have not been reported. Note that steps 9 and 13 are catalyzed by the same enzyme. Abbreviations: R5P, ribose 5'-phosphate; PRPP, 5'-phosphoribosyl pyrophosphate; PRA, 5'-phosphoribosylamine; GAR, 5'-phosphoribosylglycineamide; FGAR, 2-formamido-*N*-ribosylacetamide 5'-phosphate; FGAM, 2-formamino-*N*-ribosylacetamidine 5'-phosphate; AIR, 5-amino-1-ribosylimidazole 5'-phosphate; AICR, 5'-phosphoribosyl-5-amino imidazole; SAICAR, *N*-(5-amino-1-ribosyl-4-imidazolylcarbonyl)-L-aspartic acid 5'-phosphate; AICAR, 5-amino-1-ribosyl-4-imidazole-carboxamine 5'-phosphate; FAICAR, 5'-phosphoribosyl-4-carboxamide-5-formamidoimidazole; IMP, inosine monophosphate; XMP, xanthine monophosphate; AMPS, adenylosuccinate; AMP, adenosine monophosphate. (Data adapted from Irwin *et al.*, 1979.)

a bacterial biosynthetic pathway. Mutants have been shown to exist for 9 of the 12 individual steps leading to the synthesis of AMP. The mutants isolated show the properties of single-gene mutations in structural genes: They complement one another, are recessive in hybrid combinations, accumulate the appropriate intermediates, and lack (in most cases) a single enzymatic activity. An exception (Ade$^-$I) is defective in two steps of the pathway, which are believed to be carried out by the same enzyme, adenylosuccinate lyase (Irwin et al., 1979).

Two other most interesting exceptions to this general pattern may constitute evidence for regulatory mutations in the purine synthetic pathway. The mutant Ade$^-$P$_{AB}$ has lost *two* enzymatic activities: formylglycinamidine ribonucleotide (FGAM) synthetase (reaction 5) and glutamine-dependent amidophosphoribosyltransferse (reaction 2). These steps (Fig. 3.2) are nonsequential, and Ade$^-$P$_{AB}$ will complement neither Ade$^-$A nor ADE$^-$B. The mutation could be the result of a mutation in a single multifunctional enzyme complex which determines both activities. Alternatively, the mutation could be in a regulatory gene governing the two activities (Oates et al., 1980).

Patterson et al. (1981) have isolated another purine-requiring double mutant for Ade$^-$C (phosphoribosylglycinamide synthetase) and Ade$^-$G (phosphoribosylaminoimidazole synthetase). This mutant, Ade$^-$P$_{CG}$, could also be interpreted as evidence for a multifunctional polypeptide or a regulatory mutation in a gene governing the expression of *two* independent enzymes.

B. Pyrimidine Biosynthetic Mutants

These mutants have been isolated and characterized in mammalian cells by using the reverse selective method (Patterson and Carnwright, 1977). Furthermore, an interesting class of enzyme overproducers can be isolated by using certain pyrimidine analogs (Kempe et al., 1976). Finally, the human genetic disorder orotic aciduria results from an auxotrophic pyrimidine mutation and can be treated by adding uridine to the diet. This disorder is the result of a mutation in one of the two enzymes converting orotic acid to UMP (Worthy et al., 1974; see Fig. 3.3).

A class of uridine-requiring mutants designated Urd$^-$A have diminished activity of the first three enzymes of pyrimidine biosynthesis: aspartate transcarbamylase (EC 2.1.3.2), dihydro-orotase (EC 3.5.2.3), and carbamoylphosphate synthetase (EC 6.3.5.5) (Patterson and Carnwright, 1977; steps 1, 2, and 3; Fig. 3.3). These observations could be explained by (1) a defect in one of the structural genes causing an interference with the functioning of the other two enzymes; (2) the existence of coordinate regulation, i.e., an operon complex; or (3) a single, multifunctional polypeptide carrying information for all three enzymes. Further studies by Davidson and Patterson (1979) using an antibody precipitation technique have drawn those authors to favor the third hypothesis—that the three

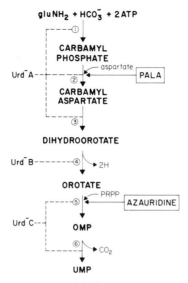

Fig. 3.3. A simplified representation of the first six steps of pyrimidine biosynthesis (Patterson and Carnwright, 1977). Abbreviations: OMP, orotidine monophosphate; UMP, uridine monophosphate.

enzymes reside on a single, multifunctional, 220,000-dalton polypeptide. A potent and specific inhibitor of aspartate transcarbamylase, used to select overproducers for this complex (Kempe *et al.*, 1976) is N-phosphonacetyl-L-aspartate (PALA). Variants resistant to PALA occur with a mutation rate of 2 to 5 $\times$ 10^{-5} per cell generation, and they owe their phenotype to the same sort of genetic redundancy previously discussed in the case of aminopterin resistance (Wahl *et al.*, 1979). The three enzymes, which copurify as a complex, are all elevated in resistant mutants, and the degree of elevation is proportional to the degree of resistance. Studies with radioactive PALA demonstrate that the increase in activity is proportional to the increase in the number of sites that bind this analog, and other physicochemical properties of the mutant enzyme are not altered. In a series of ten PALA-resistant mutants, the degree of amplification of the DNA was found to be approximately equal to the increase in messenger RNA and protein production of the complex. It may be that overproducing variants in mammalian cells frequently result from an increase in the number of genes for the protein in question, but nucleic acid hybridization studies will be required to establish this point. Another likely candidate for this status is the U complex: the enzymes orotidine-5'-phosphate decarboxylase and orotate phosphoribosyltransferase (Fig. 3.3; steps 6 and 5). Augmentation in the levels of these enzymes can be achieved by selection with pyrazofurin and azauridine, and increases up to 67-fold that of wild type have been reported (Suttle and Stark, 1979). Conversely, mutants lacking both these enzymatic activities can be selected with 5-fluorouracil by taking advantage of the fact that phosphoribosylation is necessary in order for the analog to manifest its toxicity (Patterson, 1980). This observation, coupled with the fact that most patients with the autosomal recessive genetic disorder lack both enzymatic activities (Worthy et al., 1974), reinforces the belief that these enzymes constitute a multifunctional complex.

C. Amino Acid Requirers

Mammalian cells in culture are auxotrophic for approximately 13 of the 20 or so naturally occurring amino acids. Those amino acids for which they are prototrophic are synthesized from other amino acids in one step; therefore, we would not expect to find coordinate pathways regulated in an operon-like fashion. Furthermore, because of the differences in amino acid metabolism in different tissues, one might expect to see these variations reflected in cells from these tissues cultivated *in vitro*. Both of these predictions have been realized.

Puck and his many collaborators have published an extensive series of papers that represent the pioneering studies in this area (Puck, 1975). These authors have biochemically and genetically characterized a large series of mutants that have been isolated through reverse selection and that require glycine, serine, and proline. The results of these studies have shown that the variants behave as mutations in the DNA of structural genes coding for the particular enzymes. Table 3.2 gives the properties of some of these and other auxotrophic variants.

Numerous cell lines of neoplastic origin have been shown to require asparagine because of a complete absence of the enzyme asparagine synthetase which catalyzes the formation of asparagine from aspartic acid and glutamate (Patterson and Orr, 1967). This requirement has served as the basis of a rational chemotherapeutic approach to the treatment of certain neoplasms by utilizing the enzyme asparaginase (Fig. 3.4). It has been particularly successful in the treatment of acute leukemia, in which a high proportion of cases show partial or complete remission (Hill et al., 1967). Studies in vitro have demonstrated that nonrequiring variants can be selected from such asparagine-requiring lines in asparagine-free medium. Fluctuation tests indicate that they are preexisting in the population and that the mutation rate is of the order of 10^{-6} per cell per generation (Prickett et al., 1975; Summers and Handschumacher, 1973; Morrow, 1971). The character behaves as a recessive in somatic cell hybrids. Mutation to asparagine requirement was increased by mutagenic agents but was not decreased by a so-called "antimutagen," quinacrine-HCl (Morrow et al., 1976). Although the relationship between malignancy and asparagine requirement is not clear, asparagine-requiring cell lines have recently been isolated through the use of reverse selection. Their biochemical basis is somewhat more complex. For instance, Soderberg et al. (1977) have described a respiratory-deficient mutant in

TABLE 3.2 Properties of Some Amino Acid-Requiring Variants

Amino acid	Cell line	Enzyme affected	Comments	Reference
Glycine	CHO	Serine hydroxyl-methyltransferase	Evidence for structural gene mutation	Chasin et al. (1974)
Asparagine	DON	Asparagine synthetase	Stable, low reversion	Goldfarb et al. (1977)
Asparagine	CHO	Succinate dehydrogenase	Respiratory deficient, also requires CO_2	Soderberg et al. (1977)
Argininosuccinic acid	DON	Argininosuccinic synthetase	Naturally occurring	Carritt et al. (1977)
Asparagine	Jensen sarcoma	Asparagine synthetase	Naturally occurring	Morrow et al. (1976)
Tyrosine	Rat hepatoma	Phenylalanine hydroxylase	Isolated from tyrosine-independent line	Choo and Cotton (1977)
Alanine	CHO	Alanyl-tRNA synthetase	Glutamate–pyruvate transaminase normal	Hankinson (1976)
Arginine	Reuber hepatoma	Ornithine carbamoyltransferase	Variants can be isolated that grow in ornithine-containing medium	Niwa et al. (1979)
Valine, leucine, isoleucine	CHO	Branched-chain amino acid transferase	Substitution of respective α-keto acid	Jones and Moore (1976)

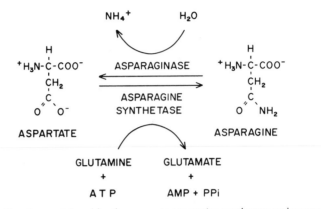

Fig. 3.4. Reactions catalyzed by the enzymes asparagine synthetase and asparaginase.

Chinese hamster cells which has a requirement for asparagine and CO_2 because of an absence of succinate dehydrogenase. Because of the uncoupling of oxidative phosphorylation, the cells also have a requirement for glucose. The biochemical basis of these enzyme-deficient mutants has been demonstrated by a number of methods, including histochemical staining. Such mutants promise to be a valuable addition to the existing repertoire of variants that occur in cultured mammalian cells.

Another class of asparagine-requiring cells has been isolated by Goldfarb *et al.* (1977), and these appear to be quite comparable to the naturally occurring neoplasms. Such variants are deficient in asparagine synthetase, are recessive in hybrid combinations, show no karyotypic abnormalities, and fail to complement one another or the asparagine-requiring Jensen sarcoma. It is notable that these variants were isolated through three successive cycles of selection. This is not surprising in view of the fact that the Jensen cell line rapidly undergoes "asparagineless death" in the absence of this amino acid (Colofiore *et al.*, 1973). Such results lead us to the optimistic conclusion that through adjustment of the experimental conditions it may be possible to isolate a wide spectrum of nutritional variants.

The model most consistent with the principle of Occam's razor would be, of course, that the asparagine requirement is the result of mutational lesions in the structural gene for the enzyme. This explanation, however, does not account for the relationship between malignancy and asparagine requirement. This correlation is certainly not a perfect one, because many neoplasms are not asparagine requiring, including revertants from asn^- to asn^+. Although we do not have information on the malignant status of the variants selected by Goldfarb *et al.* (1977), it would be surprising indeed if their tumor-producing ability were augmented. One hypothesis worth considering is that the malignant cells that are asparagine requiring have lost the ability to synthesize the enzyme through a somatic mutation that is selectively advantageous in an asparagine-rich environment. Although the anticipated selective advantage would be small, given a sufficient number of generations and the intense selective pressure occurring in a rapidly dividing tumor cell population, it might be expected that the auxotrophic cells would predominate. This hypothesis was tested by Colofiore *et al.* (1973) by observing the population dynamics of artificial mixtures of asn⁻ and asn⁺ cells. When grown in the presence of asparagine, the asn⁻ cells are at an advantage and eventually take over the population. This explains why it is not necessary to periodically purify asn⁻ lines even though nonrequiring variants are constantly arising through new mutations (see Chapter 1).

Naylor *et al.* (1976) have very carefully examined the requirements of different mammalian cell lines in culture for the essential amino acids. They have detected several unusual nutritional needs, including the inability of CHO cells to utilize cystathionine and citrulline, the inability of HTC hepatoma cells to metabolize citrulline, and the fact that KB cells can utilize homocysteine in the place of methionine. The inability to metabolize citrulline is due (in DON cells) to the absence of argininosuccinate synthetase (Carritt *et al.*, 1977), the expression of which is a dominant trait in somatic cell hybrids and appears to be controlled by a gene on human chromosome 9 (see Chapter 6 on gene mapping).

McBurney and Whitmore (1974) have isolated a cell line known as AUX B1 which is a multiple auxotroph for glycine, adenosine, and thymidine and involves a single mutation responsible for the addition of glutamate residues to

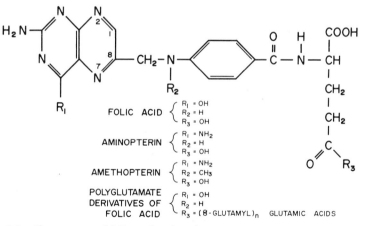

Fig. 3.5. The structure of folic acid and its derivatives. The carbon groups that may be carried at site R_2 are methyl, formyl, formimino, methenyl, and methylene. The first glutamyl residue attached at site R_3 is joined by a peptide bond to the γ-carboxyl group of the glutamate end of the folate molecule. Additional glutamyl residues become attached in tandem by γ-peptide linkages (McBurney and Whitmore, 1974).

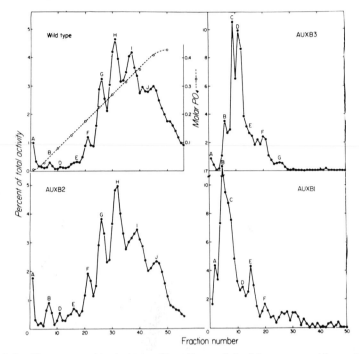

Fig. 3.6. Chromatographic elution profiles of intracellular folates extracted from wild-type and mutant cells. Exponentially growing cells were suspended in medium supplemented with ^{3}H-labeled folic acid, after which the folates were extracted and analyzed in a DEAE-cellulose column eluted with a linear phosphate buffer gradient. The following tentative identification of the folate peaks was made: A, unabsorbed folate breakdown; B, 10-formyl FH_4; C, 5-formyl FH_4 and 10-formyl FH_4Glu_2; D, 5-methyl FH_4, 5-formyl FH_4-Glu_2, and 10-formyl FH_4Glu_2; E, FH_4; F, 5-methyl FH_4Glu_2 and 5-formyl FH_4Glu_3; G, 10-formyl FH_4Glu_5, 5-methyl FH_4Glu_3, and FH_4Glu_3; H, 5-formyl FH_4Glu_5; I, 5-methyl FH_4Glu_5; J, FH_4Glu_5 (McBurney and Whitmore, 1974).

intracellular folate (Fig. 3.5). The R_2 site shown in Fig. 3.5 is the donor site for methyl groups. The mutant is in the monoglutamyl form rather than in the pentaglutamyl form (as shown by analysis by chromatography on DEAE-cellulose; Fig. 3.6). The function of polyglutamate folates is obscure, although they are thought to carry out a possible retention function preventing their leakage from the cell. In any case, the glutamyl residues are required for proper functioning of the folate, and their absence results in the auxotrophic requirement of the mutant. Another independent mutation known as AUX B3 has been mapped to the same cistron; however, the pattern of chromatographic elution (Fig. 3.5) is slightly different and such cells are *not* auxotrophic for thymidine. A temperature-sensitive AUX B1 has also been isolated. *None* of the mutants complement one another in somatic cell hybrids establishing the fact that they are all in the same cistron.

D. Carbohydrate Preferences

Differences in carbohydrate requirements of cultured cells is another potential avenue for the development of genetic variability. Studies on the carbohydrate preferences of mammalian cells go back to the 1950s (Eagle, 1959). However, such studies have been hampered by the contamination of reagents by glucose or by the production of glucose by enzymes in the serum. An extensive study by Burns *et al.* (1976) has circumvented this difficulty through the addition of catalase and glucose oxidase to the culture medium in order to eliminate false positives resulting from glucose contamination. These studies have brought to light a number of interesting differences between various cell lines in their ability to utilize various carbohydrates for growth. For instance, a hepatoma cell line can utilize various carbohydrates for growth. For instance, a hepatoma cell line can utilize any of 15 carbohydrates for growth, whereas a melanoma can utilize only 6. These differences may reflect the state of differentiation of the particular cell line. Because these "luxury functions" can be selected by nutritional modification of the medium, they should provide substantial material for the study of gene regulation.

An interesting exploitation of carbohydrate utilization has been carried out using established hamster cells by Sun *et al.* (1974). These authors have used their reverse selection procedure for the isolation of a class of mutants requiring galactose (gal$^-$). Such variants lack galactose-1-phosphate uridyltransferase (EC 2.7.7.10), the same enzyme responsible for the human inborn error of metabolism, galactosemia. Cell hybrids between these variants and human lymphocytes were found to retain only the human chromosome A2, indicating the localization of the gal$^-$ gene to A2 (Sparkes *et al.*, 1979).

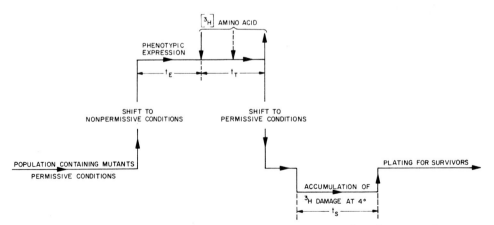

Fig. 3.7. Isolation procedure for temperature-sensitive, aminoacyl-tRNA mutants. (From Thompson *et al.*, 1975.)

TABLE 3.3 Characteristics of Some Temperature-Sensitive Growth of Animal Cells[a]

Mutant name	Cell strain	EOP np/p temperature[b]	ts phenotype	Biochemical defect	ts gene product
ts422E	BHK21,Sy hamster	$<1 \times 10^{-6}$	28 S rRNA production	rRNA processing	Ribosomal proteins?
ts546	HM-1,Sy hamster	$<1 \times 10^{-6}$	Division block	Metaphase completion	?
ts111	GM7S,Ch hamster	ND[c]	Division block	Cytokinesis	?
tsA1-S9	L, mouse	1×10^{-5}	DNA⁻	DNA synthesis	DNA ligase?
tsAF8	BHK21,Sy hamster	$<1 \times 10^{-6}$	DNA⁻	G_1 progression	?
X12	WG1A,CH hamster	$<1 \times 10^{-6}$	DNA⁻	Entry into S	?
tsB54	Mouse	$<3 \times 10^{-6}$	DNA⁻	G_1 progression	?
tsH1	CHO,Ch hamster	$<1 \times 10^{-6}$	Protein⁻	Leucine charging of tRNA	Leucyl-tRNA synthetase
ts14	V-79, Ch hamster	-1×10^{-4}	Protein⁻	60 S ribosome	Ribosomal protein?
tsAUXB1	CHO, Ch hamster	ND	DNA⁻	Folate metabolism	?
E3,H6	L6E rat myoblasts		Differentiation into myotubes	Myosin accumulation; phosphocreatine kinase increase; phos-	Control function?

		Transformed phenotype		Cellular control function
tsSV3T3	SV40-transformed 3T3 mouse cells	Transformed phenotype	Saturation density; growth on top of 3T3; lack of density inhibition of DNA synthesis; Con A agglutination; growth in low serum ± ... phorylase increase	Cellular control function
ts23A	SV40-transformed 3T3 mouse cells	Serum independence	Growth in low serum; saturation density	Serum uptake or utilization
ts-223	W-8, rat liver cell transformed by N-acetoxyacetyl-aminofluorene	Transformed phenotype	Growth in agar; morphology	ts mutation in a growth regulating function?
Chemically transformed BHK	BHK/21 Syrian hamster cells	Transformed phenotype	Growth in agar; morphology	ts mutation in a growth regulating function?

a From Basilico (1977).
b EOP, Efficiency of plating; np, nonpermissive; p, permissive.
c ND, Not determined.

E. Temperature-Sensitive Variants

A final class of variants obtained in cell culture systems bear consideration in this chapter, although not all of them represent auxotrophic gene mutations. These are conditional lethal mutations: variants that are lethal only under specific conditions. Of these, temperature-sensitive mutants are the most widely utilized. Theoretically, it should be possible to select mutants in an extremely wide range of functions using such a system. Conditional lethal mutations will enable investigators to study the cell cycle, ribosomal structure, and mutants involving complex structure–function relationships in mammalian cells.

An important class of temperature-sensitive mutants has been isolated using tritiated amino acid suicide (Fig. 3.7). In brief, cells are mutagenized, shifted to nonpermissive conditions (39°C), exposed to tritiated amino acids, and shifted to the cold for 96 hours (which enables destruction of those wild-type cells that incorporated the amino acid). They are then plated out in standard medium at 39°C and the survivors are collected and tested. By using this procedure investigators have isolated and characterized over 300 aminoacyl-tRNA synthetase mutants of Chinese hamster cells; these mutants represent seven complementation groups. All are recessive and occur in the structural genes for specific aminoacyl-tRNA synthetases. The ease with which such variants can be isolated, together with their recessive phenotype, has suggested that their respective loci are hemizygous, but none is located on the X chromosome and the question remains unresolved (Adair *et al.*, 1979; see also Chapter 1).

Basilico (1977) has extensively reviewed progress in the isolation and characterization of temperature-sensitive variants in cultured cells, and he stresses the following points:

1. Temperature-sensitive mutations probably represent single amino acid substitutions in a protein. They are generally isolated following mutagenesis, a recovery period to allow for expression, and an appropriate selection procedure in which the mutants are enriched (as, for example, Fig. 3.7).

2. Variants may be unambiguous, "good" temperature sensitives with a low reversion rate; may possess a high reversion rate; may exhibit a density-dependent growth at high (nonpermissive) temperatures; or may fail to grow at the nonpermissive temperature, but also may grow very poorly at the permissive temperature.

3. Temperature-sensitive mutants behave as recessives in somatic cell hybrids, and have been assigned to at least 20 complementation groups. The problem of how a recessive temperature-sensitive mutation might be selected in a diploid cell is as yet unresolved but certainly suggests Siminovitch's (1976) model of extensive functional haploidy in permanent cell lines.

4. Mutations affecting a variety of processes have been characterized, including cell division, DNA synthesis, RNA synthesis and metabolism, protein syn-

thesis, metabolic interconversions, and specialized functions (i.e., transformation, myotube formation). Table 3.3 summarizes some of these studies.

IV. CONCLUSIONS

Studies on auxotrophic and temperature-conditional lethals in mammalian cells have provided us with some of the most direct evidence available for mutations in structural genes as the basis of variation in mammalian cells. They have not brought to light epigenetic changes nor have they told us very much about the processes of gene regulation and control in eukaryotic organisms. This suggests that "housekeeping" functions may not be regulated, at least not in a manner that is accessible to the approaches outlined here. The fact that regulatory mutations for extended metabolic pathways, such as purine synthesis, have not been established suggests either that they may not exist or that present techniques have not yet detected them. The extremely interesting mutants described by Patterson *et al.* (1981) may indeed represent such regulatory mutations; however, alternative explanations exist. This paucity does not reflect the situation in eukaryotes in general, because there is ample evidence for the existence of regulator and operator mutations in lower eukaryotes. Rather, it may be that differentiated multicellular organisms have evolved an internal milieu so stable and so homogeneous that regulation of the synthesis or degradation of low-molecular-weight metabolites would represent excess evolutionary baggage. There are, of course, obvious exceptions: (1) the synthesis of liver enzymes that are exposed to a substantial flux in the level of their substrates and (2) the process of differentiation, which is perforce subject to great demands in terms of gene regulation. These questions will be dealt with in subsequent chapters.

4

Mechanisms for the Exchange of Genetic Information in Cultured Cells

I. INTRODUCTION

The first three chapters of this volume have dealt with the problem of genetic variation in cultured cells and its specific molecular mechanisms. Although a commentary on eukaryotic genetic variability requires the consideration of the results of cell hybridization, we have not up to this time analyzed in detail the mechanisms of this process. Obviously, genetic analysis requires some mechanism by which differing sets of genetic information can be combined within the same cell. In this chapter we will discuss various means by which eukaryotic cells, in whole or fractionated components thereof, may be reassembled to give rise to viable cellular composites. The most widely utilized approach is cellular hybridization, but the uses of purified DNA, viruses, and subcellular components also deserve consideration. Each of these mechanisms is of great interest for its own sake, and each has been of value in answering different questions concerning the genetics of eukaryotic cells.

II. CELL HYBRIDIZATION

A recognition that somatic cells are capable of fusion has its roots in the classic literature of nineteenth century histology. However, the first demonstration of this phenomenon *in vitro* we owe to Barski *et al.* (1960). They observed that when two tumor-producing cell lines from the mouse were cocultivated, a third, highly malignant line arose that possessed a chromosome number additive of the two parents and exhibited marker chromosomes from both. Harris (1970) recognized the value of heterokaryons for the study of gene regulation (see Chapter 5) and Pontecorvo (1959) stressed the utility of cell culture methods for the establishment of linkage relationships. The manner in which cell hybrids were employed for gene mapping was actually somewhat different from the way Pontecorvo had envisioned. Based on his experience with gene mapping in fungi, he had suggested that linkage groups could be identified through partial haploidization. In actuality, the assignment of genes to chromosomes, or to specific regions of chromosomes, has been pursued by following the random loss of chromosomes from extremely complex allotetraploids (see McKusick, 1980, for a review up to that time, and Chapter 6 for a more recent summary).

Although the earliest investigations into cell hybridization were performed without the aid of a selective system, Littlefield (1964b) took advantage of the HAT selective system (see Chapter 2) to obtain hybrids between mouse lines deficient for TK and HPRT. Subsequently, many of the drug resistance and auxotrophic markers considered in the previous chapters have been used for the selection of cell hybrids. Thus, any two permanent cell lines with the general configuration of recessive mutations

$$A^-B^+ \times A^+B^-$$

can be hybridized in the appropriate selective medium provided the reversion rate for the auxotrophic markers is substantially lower than the rate of hybridization (Table 4.1).

A number of "half-selective" systems have also been devised; these systems take advantage of the fact that hybridization can rescue a nondividing cell type (Davidson and Ephrussi, 1965), and they utilize a biochemical marker (such as HAT selection) to eliminate a permanent, rapidly dividing cell line, which is one of the parents in a hybridization protocol. The other parent may be a diploid fibroblast line (or any nontransformed cell type) that is a slow grower and is outpaced by the rapidly proliferating hybrid.

A system for selecting hybrids without the use of genetic markers has been devised by Wright (1978). Two different cell lines are treated with lethal doses of irreversible biochemical inhibitors and subsequently fused. Only heterokaryons receive a full complement of undamaged molecules necessary for survival and give rise to functional hybrids. This method is of value in that it avoids the laborious process of introducing genetic markers into cell lines; however, it possesses the disadvantage of requiring an extremely fine balancing of toxicity in order to avoid killing the parental cells before hybridization has been achieved.

Although cell hybrids may be obtained spontaneously (Littlefield, 1964b), the standard method today uses agents that expedite the rate of fusion. Inactivated viruses, mainly Sendai virus, have been found to greatly increase the rate of

TABLE 4.1 Selective Systems for the Production of Cell Hybrids

Parental genotypes	Hybrid phenotype	Selective medium	Reference
HPRT$^-$ × TK$^-$	HPRT$^+$ TK$^+$	HAT	Littlefield (1964)
HPRT$^-$ × ASN$^-$	HPRT$^+$ ASN$^+$	HAT-deficient-asparagine	Morrow et al. (1973)
Gly$^-$ Ad$^-$ × Ad$^-$ Pro$^-$ Thy$^-$ Pro$^-$	Gly$^+$ Ad$^+$ Pro$^-$ Thy$^+$	Glycine, adenosine, and thymidine omitted	Thompson et al. (1973)
HPRT$^-$Ou^R × any wild-type cell HPRT$^+$, Ou^S	HPRT$^+$ Ou^R	HAT$^+$, ouabain	Ozer and Jha (1976)
AdeA × AdeB	Adenine independent	Adenine deficient	Kao (1973)
5-Fluoroorotic acid-resistant × Y^{-a} (uridine requiring)	Uridine independent, Y independent	Uridine deficient, Y deficient	Krooth et al. (1979)
HPRT$^-$ × APRT$^-$	HPRT$^+$ APRT$^+$	"GAMA," Guanine, adenine, mycophenolic acid,^{b} and azaserine	Liskay and Patterson (1979)

^{a} Y, Any low-molecular-weight nutritional requirement.
^{b} Prevents conversion of guanine to adenine.

fusion (Okada, 1958; Neff and Enders, 1968), enabling the generation of hybrid cells at high frequencies. Routine production of cell hybrids in large quantities without resort to biological agents has been made possible through the discovery of the fusogenic properties of polyethylene glycol (Pontecorvo, 1975). When added to tissue culture medium in concentrations ranging from 5 to 50%, this substance brings about the coalescence of cell membranes and the formation of hybrids following selection. Although yields are higher when closely confluent monolayers are fused, fusion in suspension can also be implemented. The molecular weight of the polyethylene glycol appears to be a factor in the fusion process, although different brands, time of exposure, and concentration of PEG also appear to be critical. Because it is extremely cheap, effective, and lends itself quite easily to standardization, this now appears to be the method of choice for cell fusion (Davidson and Gerald, 1976). Figure 4.1 shows the rather striking appearance of a culture of cells immediately following removal of PEG.

Although the early experiments by Barski *et al.* (1960) were carried out using cells of the same species, it soon became evident that cells from different species could combine to form viable hybrids and that these hybrids subsequently segregate chromosomes according to rates that were determined by the species involved in the hybridization and the division time of the parental cells. A striking example of the role of parental division time in the determination of chromosome segregation patterns has been provided by Minna and Coon (1974). These authors demonstrated that although human–rodent hybrids ordinarily segregate human chromosomes, the pattern is reversed when rapidly dividing permanent human cells are fused with freshly isolated mouse cells. Thus, the direction of chromosome loss is determined by properties associated with relative growth rates of the parental cell types rather than by the species of the parental types. A typical cellular hybridization (Figs. 4.2 and 4.3) yields cells that have chromosomes contributed by both parents and a cellular morphology that is mainly distinguished by the larger size of the hybrids.

In general, chromosome loss is more rapid during the early period following fusion and then tends to stabilize. Chromosome loss appears to be nonrandom, and the mechanism underlying this phenomenon is not understood at present (Kucherlapati and Ruddle, 1976).

Our inability to control the rate of direction of chromosome loss in hybrids has occasioned a search for specific mechanisms by which this process might be influenced. Pontecorvo (1971) has developed a method by which one parent is treated with X irradiation or with BUdR followed by visible light. Such treatments result in the preferential expulsion from the hybrids of the chromosomes from the treated parent.

In the HAT selection system, cell hybrids are selected by virtue of the fact that certain metabolic enzymes (HPRT and TK) produced by alternative parents are both synthesized in the hybrid. Thus, these ''housekeeping'' functions are ex-

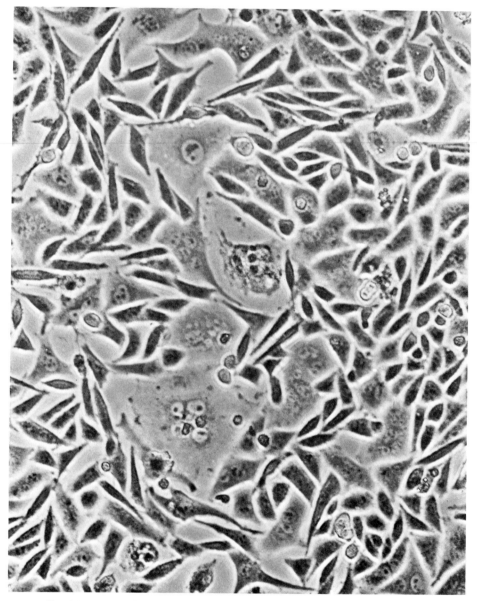

Fig. 4.1. Fusion of a mixed culture consisting of rat sertoli cells and mouse fibroblasts 1 hour after treatment with PEG 1000. Note numerous multinucleate cells. (Photo courtesy of Dr. James Hutson.)

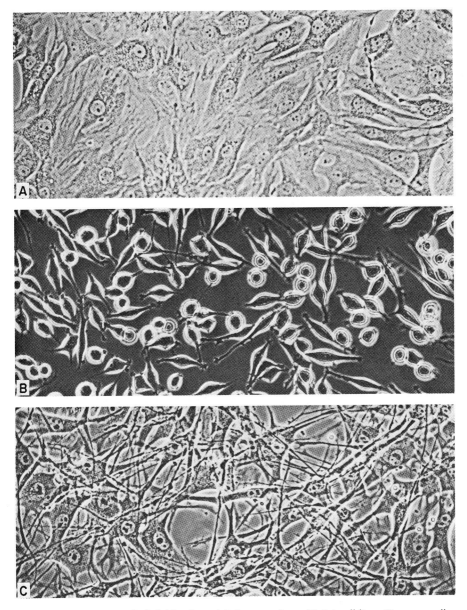

Fig. 4.2. Interspecific hybrid cells and their parent lines. (A) Rat cell line; (B) mouse cell line; (C) cell hybrid. Note intermediate morphology of hybrid line (Ephrussi and Weiss, 1969).

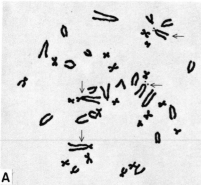

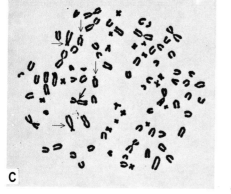

Fig. 4.3. Chromosomes of rat, mouse, and rat–mouse hybrid (A, B, and C, respectively) shown in the previous figure. Arrows indicate marker chromosomes from the parental lines appearing in the hybrid (Ephrussi and Weiss, 1969).

pressed by both genomes, which has proved to be a general rule for hybrids (Table 4.1; Davidson, 1973). In numerous instances, gene products from the parental species will recombine in the cytoplasm to form interspecific hybrid proteins. This contrasts with the case of differentiated characters, which are so often suppressed in cell hybrids (Chapters 7, 8, and 9).

III. GENETIC EXCHANGE USING CELL COMPONENTS

An exciting area of investigation that has developed takes advantage of the ability of cells to be separated into viable nuclei ("karyoplasts") and cytoplasms ("cytoplasts") and to be restored by using fusion techniques. Cytochalasin B (Fig. 4.4) is a metabolite of the fungus *Helminthosporium*, and at a concentration of 10 μg/ml at 37°C it causes nuclear extrusion. Upon replacement with normal medium, nuclei are reabsorbed without apparent ill effect. These nuclei can be subsequently purified by the use of centrifugation, which shears off the nuclei from the stalk that connects them to their respective cytoplasms. Such fragments are viable for approximately 48 hours and during this period can be reassembled using PEG-mediated fusion. The mode of action of cytochalasin B is obscure, but it is believed to cause the disassembly of cytoplasmic microfilaments. Separated cytoplasts and karyoplasts appear normal under the electron microscope. However, they will, when left to their own devices, fail to regenerate. When karyoplasts and cytoplasts are fused with PEG, the resulting product is an apparently normal cell.

Thus, if cytoplasts from a chloramphenicol-resistant cell are fused with ouabain-resistant, thioguanine-resistant cells and the products are plated in ouabain and CAP, only cybrids will be obtained. The unfused ouabain-resistant cells will be killed by the CAP, and the unenucleated CAP-resistant cells will be killed by

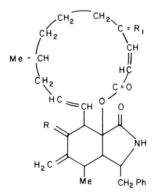

Fig. 4.4. Structure of cytochalasin B.

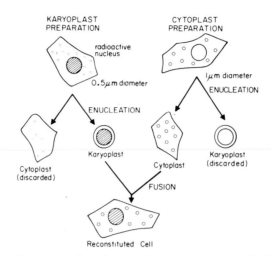

Fig. 4.5. Schematic representation of preparation of reconstituted cells. Karyoplasts were derived from cells labeled with [³H]thymidine and small (0.5 μm diameter) latex spheres and fused to cytoplasts from cells labeled only with large (1.0 μm diameter) latex spheres. The latex spheres permitted recognition of the source of the cytoplasm. Mononucleated cells with radio-activated nuclei and only large latex spheres in the cytoplasm were considered as cells reconstructed from components derived from different sources. Fusions resulting from whole cell contaminants in the karyoplast preparation were recognized by the numerous small spheres in the cytoplasm and were excluded from consideration as reconstructed cells. (From Veomett *et al.*, 1974.)

the ouabain. Resistant clones can be double-checked to rule out a possible cell hybridization by testing HAT medium. A rare CAP-resistant thioguanine-resistant, ouabain-resistant cell hybrid would live in HAT medium, in addition to possessing a hybrid chromosomal constitution. Another approach used in the recognition of reconstituted cells has been to label cells with polystyrene beads of different sizes which allows one to follow the different cell components during the reassembly procedure (Veomett *et al.*, 1974; Fig. 4.5).

Using this technique, it has been shown that chloramphenicol (Wallace *et al.*, 1975), oligomycin (Breen and Scheffler, 1980), erythromycin (Doerson and Stanbridge, 1979), and antimycin D resistance (Harris, 1978) are cytoplasmically determined characters. This conclusion is based on the fact that for these four resistant types cytoplasts alone are sufficient to carry the resistant factor when fused with sensitive cells.

IV. TRANSFER OF GENETIC INFORMATION USING PURIFIED METAPHASE CHROMOSOMES

Because chromosomes are intact organelles that carry genetic information and that have centromeres and attendant architectual proteins, they should be able to avoid extra- and intracellular degradation of their information-carrying capacity and successfully achieve stabilization and expression in a foreign environment. Although it was well documented that radiolabeled chromosomes reach the nucleus, no firm evidence was available for expression of such foreign genetic information in the host cytoplasm (Chorazy *et al.*, 1963). This lacuna has been filled in recent years (Burch and McBride, 1975; Willecke and Ruddle, 1975; McBride and Ozer, 1973; Wullems *et al.*, 1975, 1976a,b). These experiments utilized interspecific crosses in which the identity of the donor gene product (usually HPRT) could be unambiguously demonstrated by biochemical means. In some cases, the donor marker appeared to be stably integrated, whereas in other clones this was not the case. In none of these hybrids could intact chromosomes from the donor species be recognized, suggesting that fragmentation into small pieces had occurred. Even the cotransfer of the thymidine kinase and galactokinase genes was accompanied by fragmentation (Wullems *et al.*, 1977; McBride *et al.*, 1978). The frequency of transfer of any gene by chromosome-mediated transfer is approximately 10^{-7} (McBride and Athwal, 1976).

Although the exact basis for the unstable state of transferants is not known at this time, Degnen *et al.* (1976) have observed that in the case of intraspecific gene transfer the transferred gene is unstable and stabilization occurs at the rate of 10^{-5} per cell per generation. Thus, when large populations of cells are maintained for extended periods in selective medium, the line will eventually stabilize through selection.

The frequency of transformation can be greatly augmented through the use of chemical protective agents. Miller and Ruddle (1978) demonstrated that when HeLa S3 chromosomes were added to HPRT$^-$ mouse A9 cells, up to 4×10^{-5} "transgenotes" per cell could be obtained when the chromosomes were suspended in $CaCl_2$ and the cultures were treated briefly with dimethyl sulfoxide (DMS). In addition, the cotransfer of syntenic (on the same chromosomes) markers was facilitated and, in some cases, the presence of human chromosomal material could be demonstrated cytologically (Fig. 4.6). In this report, as in other cases, clones are a mixture of stable and unstable transgenotes.

A subsequent report from Ruddle's laboratory (Scangos *et al.*, 1979) further characterized the transferance process through the use of a probe for the transferred thymidine kinase gene. Chromosomes were extracted from an LMTK$^-$ cell line that carried the herpes simplex TK gene in an integrated state and were used to transform LMTK$^-$ cells lacking the virus. In four independent transformants expressing viral TK, it was found that the gene was at least 17 kb in size and that it stabilized at lower numbers of genes per cell following prolonged cultivation in HAT medium. The site of integration of the TK gene, however, is not at the site of the homologous defect in the host cell, as shown by gene mapping experiments of transgenotes. This is true at least in the case of the human TK gene transferred into LMTK$^-$ cells (Willecke *et al.*, 1978).

Fournier and Ruddle (1977) have reported the use of a promising technique:

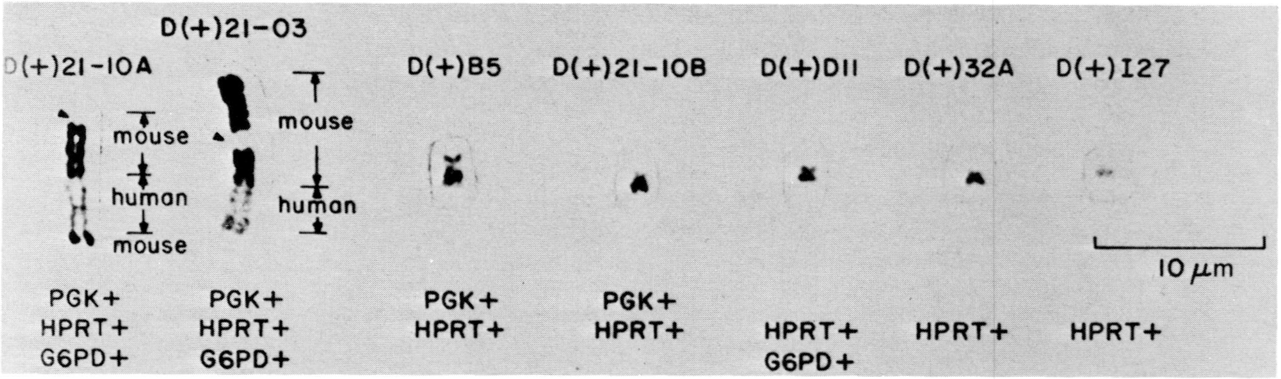

Fig. 4.6. Chromosome banding patterns in different transformed cell lines demonstrated with alkaline Giemsa. The first two chromosomes contain pale-staining human material and dark-staining mouse material in the arms with characteristically pale mouse centromere regions (—). The remaining chromosomes consist of human-staining material only. Human X-linked isozymes expressed by each cell line are noted. (From Miller and Ruddle, 1978.)

so-called "microcell fusion." Cells in logarithmic phase are treated for 2 days with colcemid and enucleated with cytochalasin B. This results in formation of microkaryoplasts with one or a few chromosomes. By this technique intact chromosomes are transferred and linked genetic markers coexpressed. Klobutcher and Ruddle (1979) have used this approach to transfer chromosomes from HeLa S3 cells to mouse LMTK⁻ cells. Not only was the expression of the TK gene followed but also the expression of two other genes, galactokinase and procollagen I. The genetic data combined with cytogenetic analysis of the transferred segment ("transgenome") indicates that the gene order is centromere–galactokinase–(TK–procollagen I) on the long arm of human chromosome 17.

V. TRANSFORMATION USING PURIFIED DNA PREPARATIONS

The use of purified preparations of DNA would certainly be the most direct mechanism by which genetic information could be transferred between cells. First demonstrated in *Pneumococcus* (Avery *et al.*, 1944), it was subsequently investigated in a number of microorganisms, including tissue culture cells. However, early eukaryotic cell transformation experiments were not conclusive, and the development of cellular hybridization overshadowed transformation as a means of transferring hereditary information.

In the past 3 years, however, the problem has been reinvestigated with encouraging results because of improvements in the technology of DNA purification and the protection of the donor DNA through the use of calcium phosphate similar to the procedure employed in metaphase chromosome transfer. This has been demonstrated by Wigler and his collaborators, who showed that mouse cells deficient in APRT could be transformed with DNA prepared from hamster, human, and mouse cells. The transformants were both stable and unstable, a condition reminiscent of the situation in the case of chromosome transfer. Transformants were produced at a frequency of up to 10^{-5} per cell (Wigler *et al.*, 1979a). A subsequent report by the same laboratory (Wigler *et al.*, 1979b) described the results of experiments in which TK⁻ cells were transformed with DNA carrying the thymidine kinase gene of herpes simplex in purified form. It is striking to note that cotransformation with highly characterized DNAs from a variety of sources (φX174, plasmid pBR322, or globin genes from the rabbit) was extremely successful. Such a high index of cotransformation indicates that a tiny fraction of the recipient cells are in a state of competency and that uptake of a selectable DNA enables the identification of those cells that are also capable of assimilating simultaneously other types of DNA. This observation should prove to be extremely useful for the investigation of the genetics of a variety of unselectable characters (such as those regulating differentiated functions). In fact, cotransformation has been attempted by Hsiung *et al.* (1980) using purified HSV DNA and human β-globin sequences. It was shown that the human β-globin genes could be introduced into the cells; however, no globin messenger could be detected.

An application of the techniques of transformation to intact animals has been successfully executed by Cline *et al.* (1980). These workers isolated DNA from methotrexate-resistant (see Chapter 2) mouse 3T3 cells and used this preparation to transform isolated marrow cells. These marrow cells were then injected into irradiated mice that were subsequently treated with methotrexate. The results of these experiments (which utilized a distinctive chromosomal marker in the transformed cells) established that *in vivo* treatment of host animals (with methotrexate) will enrich for the transformed cell type.

VI. TRANSFER OF INFORMATION THROUGH DIRECT MICROINJECTION OF RNA AND DNA

Finally, it is possible to obtain uptake and expression of both RNA and DNA through direct microinjection. Although much microsurgery of this sort has been carried out in amphibian eggs (Mertz and Gurdon, 1977), mammalian cells are smaller than *Xenopus* oocytes by a factor of 10^{-5}, and it was therefore necessary to develop procedures utilizing glass pipettes for this purpose (Diacumakos, 1973). Capecchi (1980) has now shown that LMTK$^-$ cells will express the DNA from the herpes simplex virus when it is introduced directly into the nucleus. The injected marker is stable in nonselective medium, and the proportion of cells expressing the gene is so high (50–100%) that the technique should lend itself to the study of differentiation and to other questions in which it is required that an entire population of donor cells be transformed.

In the case of mRNA injection, Liu *et al.* (1979) have established that the microinjection of 10^2 cells will allow the synthesis of products that can be detected. Mouse LMTK$^-$ cells were shown to synthesize human fibroblast interferon, TK, HPRT, APRT, and propionyl-CoA carboxylase in response to injected RNA. Furthermore, by using partially purified mRNA fractions it was possible to show a specificity for interferon synthesis and to assign its production to 1 of 10 fractions of messenger.

VII. CONCLUSION

The methods available for the combining of genetic information from different sources have made possible a wide variety of experimental approaches, especially in the last 5 years. Thus, it is possible to localize and map genes using interspecific cellular hybridization and to study the problem of regulation on a cellular level with the aid of heterokaryons. Other procedures make it possible to investigate the integration of genetic information and the expression of viral and prokaryotic genes in eukaryotes, such as the herpes simplex thymidine kinase gene introduced into BUdR-resistant mouse cells. The present widespread availability of recombinant DNA technology allows for the production of highly purified genes from a variety of sources that, in conjunction with transformation and other techniques, will yield answers to questions such as:

1. What is the molecular nature of the signals by which tissue-specific functions are regulated in the course of differentiation?

2. How are fragments of chromosomal material integrated into host cells and what is the relation of this process to normal chromosomal structure?

3. Can these methods be employed for "genetic engineering"? More specifically, Can defective cells be removed from the body, treated appropriately, and replaced in a functional state in numbers sufficient to approximate a healthy phenotype?

4. What is the relationship between gene expression and chromosomal location?

5

The Regulation of Gene Expression in Heterokaryons

I. INTRODUCTION

By utilizing the technique of cell fusion, investigators can now study the problem of gene regulation in eukaryotic organisms in a direct and innovative manner. Previously we have discussed the techniques by which cells can be made to fuse and coalesce, resulting in the sharing by different nuclei of a common cytoplasm. By combining cells that differ in certain important biological properties, it is possible to ask questions about dominance of control processes, nucleocytoplasmic interactions, and gene complementation. Certain questions, such as the mapping and dissection of regulatory functions by genetic analysis, may be answered through the use of stable, proliferating hybrids that segregate chromosomes. However, issues of great interest, such as the mechanism of activation and suppression of genetic messages, can best be resolved by the use of heterokaryons: short-term combinations in which nuclear fusion has either not yet occurred or is prevented through the use of chemicals or irradiation. Because it is possible to generate a large, unselected population of independent fusion events, gene interactions can be studied immediately and without the possibility of chromosome loss. In fact, several important questions, such as the fluidity of membrane components and the events involved in the activation of the nucleus, have been investigated using cell heterokaryons (Harris, 1970).

Most of the data that will be discussed here are obtained by using interspecific heterokaryons: multinucleate cells formed by the fusion of cell types from different species (Fig. 5.1). In such combinations genes from both sets of parents actively transcribe RNA. This surprising finding indicates that no intracellular mechanisms for the recognition of interspecific incompatibility exist. This obser-

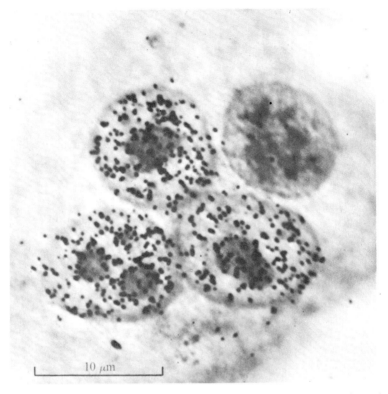

Fig. 5.1. Autoradiograph of a tetranucleate cell containing three HeLa nuclei and one Ehrlich nucleus. The HeLa cells had been grown in tritiated thymidine before the heterokaryons were produced. The HeLa nuclei are labeled and the Ehrlich nucleus is not. (From Harris, 1970.)

vation applies even to such distantly related crosses as HeLa cells and tobacco protoplasts, which will form heterokaryons that appear to be viable, at least for short periods (Jones *et al.*, 1976).

II. NUCLEAR REACTIVATION IN HETEROKARYONS

A considerable volume of literature has emerged from the extensive studies of Harris and his co-workers. This work has been reviewed in detail (Harris, 1970); here we will discuss the principal findings and relate them to other investigations utilizing heterokaryons.

When two morphologically or radioisotopically marked cell types are mixed together in the presence of inactivated Sendai virus, a coalescence characterized by reorganization of cytological architecture takes place, resulting in a redistribution of membrane components. In human–mouse heterokaryons, the surface antigens become mixed after fusion (Watkins and Grace, 1967). By using double-fluorescent antibody-labeling techniques that detect both mouse and human antigens, Frye and Edidin (1970) have shown that this process requires approximately 40 minutes. These studies provide support for the concept of membrane fluidity and the mosaic model of membrane structure (Singer and Nicolson, 1972).

Heterokaryons have also proved useful in the study of the regulation of the cell cycle. When cells in different stages of the cell cycle are fused, in some cases signals emanating from one nucleus can induce activation in the other nucleus. For instance, when S phase cells are fused with G_1 or G_0 cells, DNA synthesis is rapidly induced. This induction is due to proteins that move from the S nucleus to the G_1 or G_0 nucleus (Graves, 1972). On the other hand, G_1 or S nuclei inhibit the progression of G_2 nuclei into M until DNA replication is completed (Rao and Johnson, 1974). A phenomenon referred to as premature chromosome condensation is brought about when an interphase cell is fused with a mitotic cell. The morphology of the prematurely condensed chromosome is a function of the stage of interphase in the G_1 cell at the time of fusion. Premature chromosome condensation in the S phase results in uneven chromosome condensation and a pulverization of the chromosomes (Unakul et al., 1973).

In order to study the control of macromolecular synthesis, the fusion of various differentiated cells was carried out. In Harris's pioneering studies, rabbit macrophages, rat lymphocytes, and hen erythrocytes were fused with HeLa cells; this technique enabled Harris to investigate the regulation of nucleic acid and protein synthesis. The most important finding from these studies is that in heterokaryon combinations reactivation of macromolecular synthesis occurs (see Fig. 5.2; Table 5.1) when quiescent and active cells are fused. When erythrocytes are fused with HeLa cells, the erythrocyte nuclei are divested of their cytoplasm and respond to foreign cytoplasmic signals. Thus, there must be a general signal that emanates from the HeLa cell and that activates other species and other cell types.

The reactivation of DNA synthesis in a previously dormant nucleus appears to be a generally demonstrable phenomenon, because Gurdon (1967) showed that nuclei from adult frog brains transplanted to enucleated eggs respond by activation of DNA synthesis. Similarly, Jacobson (1969) has shown that in heterokaryons between mouse neurons and monkey fibroblasts, DNA synthesis is activated in the neuronal nuclei. In the case of the hen erythrocyte–HeLa cell heterokaryons, a drastic reorganization takes place in which the hen nucleus becomes derepressed, as shown by the increase in nuclear volume and dry mass (Fig. 5.3). Furthermore, there is a dispersion of chromatin and increased affinity to intercalating dyes such as acridine orange and ethidium bromide (Bolund et al., 1969). These substances, which sandwich into DNA, do so with greater

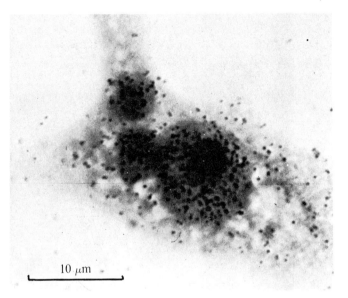

10 μm

Fig. 5.2. Autoradiograph of a heterokaryon containing one HeLa nucleus and two chick erythrocyte nuclei exposed for 4 hours to a radioactive RNA precursor. The HeLa nucleus and the erythrocyte nuclei are synthesizing RNA (Harris, 1970, p. 36). The general principles that emerge from this work are that RNA synthesis is "dominant" over nonsynthesis and that DNA synthesis is "dominant" over nonsynthesis (see Table 5.1). (From Harris, 1970.)

TABLE 5.1   Synthesis of RNA and DNA in Heterokaryons[a,b]

	RNA	DNA
Cell type		
HeLa	+	+
Rabbit macrophage	+	0
Rat lymphocyte	+	0
Hen erythrocyte	0	0
Cell combination in heterokaryon		
HeLa–HeLa	+ +	+ +
HeLa–rabbit macrophage	+ +	+ +
HeLa–rat lymphocyte	+ +	+ +
HeLa–hen erythrocyte	+ +	+ +
Rabbit macrophage–rabbit macrophage	+ +	0 0
Rabbit macrophage–rat lymphocyte	+ +	0 0
Rabbit macrophage–hen erythrocyte	+ +	0 0

[a] From Harris (1970).

[b] 0, No synthesis in any nuclei; 00, no synthesis in any nuclei of either type; +, synthesis in some or all nuclei; + +, synthesis in some or all nuclei of both types.

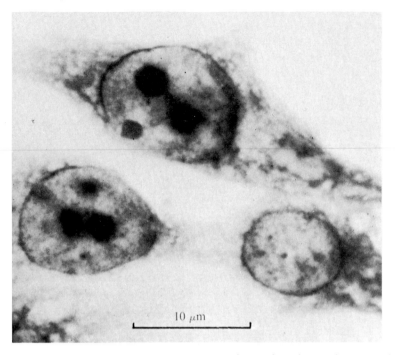

Fig. 5.3.   A heterokaryon containing one HeLa nucleus and one hen erythrocyte nucleus which has been reactivated. The erythrocyte nucleus is now greatly enlarged and the chromatin within it is much less condensed. (From Harris, 1970.)

facility when the DNA is more attenuated, which is indicative of gene activation. Following these events, there is a migration of human nucleospecific proteins into the chick nucleus. These proteins have been analyzed by isolating nuclei from chick–HeLa heterokaryons and then separating the chick from the mammalian nuclei on sucrose gradients. By using this approach investigators have shown that a nonhistone chromosomal protein with a molecular weight of 50,000 is selectively accumulated. On the other hand, enzymes and antigens characteristic of the mammalian cytoplasm are selectively excluded from the erythrocyte nucleus.

III. THE EXPRESSION OF GENETIC INFORMATION IN HETEROKARYONS

Interspecific heterokaryons provide an opportunity to analyze production of specific RNAs and proteins from an erythrocyte nucleus that was previously completely repressed. Surface antigens of different specificity are contributed by

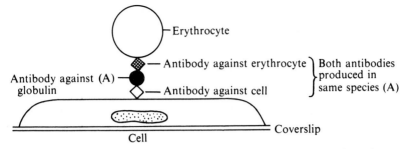

Fig. 5.4. Diagrammatic representation of the technique of mixed immune hemadsorption, a mixed antiglobulin technique. Adsorption of the sensitized erythrocyte indicates the presence of the species-specific antigen on the surface of the cell. (From Harris, 1970.)

the parents and can be investigated through the use of immune hemadsorption (Fig. 5.4). When chick erythrocytes are fused with mouse or human cells, hen-specific antigens are present at first and then disappear.

This removal of these previously synthesized substances is an active process in which human or mouse antigens continue to be produced (Fig. 5.5). After approximately 4 days, the heterokaryons are still intact; the mammalian cell was irradiated with X rays prior to fusion. In such fusion products, the reappearance of the antigen is tied to the development of nucleoli in the hen nuclei (Figs. 5.6 and 5.7).

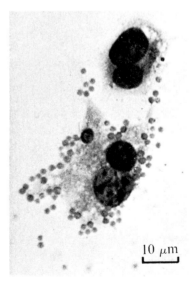

Fig. 5.5. A heterokaryon containing two mouse nuclei and one hen erythrocyte nucleus 18 hours after cell fusion. The hemadsorption reaction reveals the presence of hen-specific antigens on the surface of the cell (Harris, 1970).

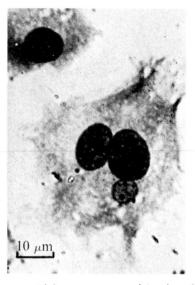

Fig. 5.6. A heterokaryon containing two mouse nuclei and one hen erythrocyte nucleus 5 days after cell fusion. The erythrocyte nucleus has been reactivated, but the absence of any hemadsorption shows that the hen-specific antigens are no longer present on the surface of the cell. The erythrocyte nucleus shows a small nucleolus (Harris, 1970).

An alternate approach to this problem is to utilize the hypoxanthine guanine phosphoribosyltransferase-deficient mouse A9 L-cell line (Chapter 2). When such cells are fused with chick erythrocytes, the same pattern of expression of enzyme and nucleoli occurs: an increase in visible nucleoli is correlated with the appearance of the enzyme. Therefore, two unrelated gene products are coordinate in their expression (Fig. 5.8).

Although the efforts of Harris and his collaborators clearly show that "housekeeping" or nonspecialized functions can be activated in heterokaryons, the situation with regard to "luxury" (differentiated) functions appears to be quite different. Alter *et al.* (1977) have shown that in heterokaryons between mouse erythroleukemia cells and human fibroblasts only mouse globin chains and mouse globin messengers are produced. This is most likely a transient situation because cell hybrids between the same cell types actively synthesize human hemoglobin (Deisseroth and Hendrick, 1979). This finding suggests that the activation of the human globin genes requires a message that cannot pass through the nuclear membrane, although it is diffusible from the mouse to the human genome when the two genomes reside within the same nucleus. If this is the case, then the situation encountered by Harris's group in the previously described studies on reactivation of macromolecular synthesis must involve a different level of regulation, perhaps a "rough" as opposed to a "fine" tuning mechanism.

Similar findings have been published by Rechsteiner and Hill (1975). When these authors fused a mouse cell line that cannot utilize nicotinic acid with a human line that can, the heterokaryons synthesized and formed nicotinic acid at a rate comparable to the human parental cell, as demonstrated by autoradiographic grain counts. However, after hybrids had developed from the heterokaryons, synthesis was suppressed to 1/10 that of the human parent. The authors suggest that either repression requires a round of nucleic acid synthesis or that there are repressors that are restricted to the nucleus. They rule out other trivial explanations, such as chromosome loss from the hybrids, and mention a similar system of regulation that has been described in *Neurospora* (Burton and Metzenburg, 1972).

Furthermore, it should be mentioned that Szpirer (1974) has made comparable observations in the case of albumin synthesis, which is not activated in chick nuclei following fusion with rat hepatoma cells. Finally, chick myosin was not produced in chick erythrocyte–rat myoblast heterokaryons (Carlsson *et al.*, 1974a).

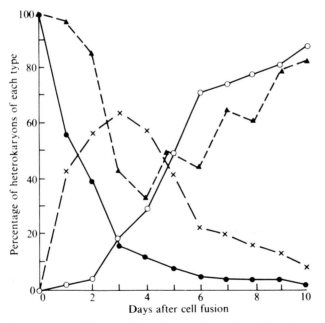

Fig. 5.7. Reappearance of hen-specific antigens on the surface of heterokaryons made by fusing irradiated mouse cells with chick embryo erythrocytes. The relationships between the enlargement of the erythrocyte nucleus, the development of nucleoli within it, and the disappearance and reappearance of the hen-specific surface antigens are shown. ●, Heterokaryons with unenlarged erythrocyte nuclei; ×, heterokaryons with enlarged erythrocyte nuclei but no visible nucleoli; ▲, heterokaryons with enlarged erythrocyte nuclei containing visible nucleoli; ○, heterokaryons showing hen-specific surface antigens (Harris, 1970).

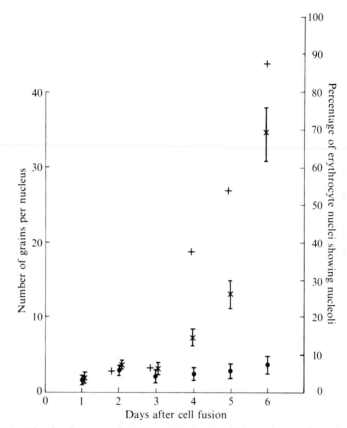

Fig. 5.8. The development of HPRT activity in A₉–chick erythrocyte heterokaryons as measured by their ability to incorporate tritiated hypoxanthine into nucleic acid. The incorporation of hypoxanthine in the heterokaryons is initially only marginally greater than that in A₉ cells alone, but when the erythrocyte nuclei develop nucleoli, this incorporation increases markedly. ●, A₉ cells; ×, heterokaryons; +, erythrocyte nuclei showing nucleoli (Harris, 1970).

An exception to this general pattern is the situation encountered with tyrosine aminotransferase, whose synthesis is suppressed in heterokaryons between producing and nonproducing cells (Chapter 7). Furthermore, it has been shown in myeloma–fibroblast heterokaryons that immunoglobin production is rapidly suppressed (Chapter 9). The reasons for these exceptions to the general pattern are not clear but suggest that more than one mechanism of gene repression may be occurring in somatic cells.

IV. THE MECHANISM OF NUCLEAR ACTIVATION IN HETEROKARYONS

Some studies have been carried out in an effort to analyze the molecular mechanism of nuclear activation. It is known that RNA synthesis by the chick erythrocyte nucleus commences shortly after fusion to HeLa cells (Harris, 1967), as demonstrated by autoradiography using [³H]uridine. A technique was therefore developed to allow, by the use of density gradient centrifugation, the separation and examination of the chick and HeLa nuclei after heterokaryon formation. By reisolation of the nuclei after activation and examination of the RNA on a sucrose gradient, the pattern of RNA synthesis can be followed. Immediately following fusion, polydisperse RNA is not seen in the HeLa cell nuclei, most of the RNA being 28 S and 16 S (ribosomes) (Fig. 5.9). On the other hand, the erythrocyte nuclei initially make only polydisperse (messenger) RNA. After the nuclei are reactivated, 28 S and 16 S RNA appear. At this stage of heterokaryon formation, no chick proteins are being manufactured; this finding suggests that the synthesis of messenger must be accompanied by other molecular events in

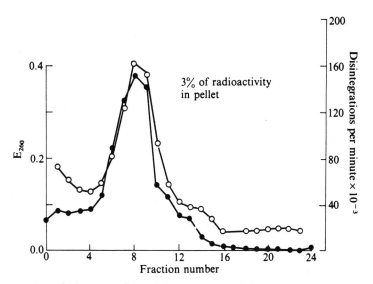

Fig. 5.9. Sucrose density gradient of the RNA extracted from HeLa nuclei isolated from HeLa–erythrocyte heterokaryons 48 hours after cell fusion. The cells were exposed to a radioactive RNA precursor for 4 hours before enucleation. The bulk of the radioactivity has been incorporated into the 28 S and 16 S RNA components. Very little radioactivity has been incorporated into "polydisperse" RNA, which, under these conditions of centrifugation, sediments to the bottom of the centrifuge tube. $\bigcirc$, E_{260}; $\bullet$, radioactivity (Harris, 1970).

TABLE 5.2   Grain Counts over Nucleoplasm and Cytoplasm in HeLa Cells in Which Either the Whole Nucleus or the Nucleolus was Irradiated[a,b]

	Number of cells	Total number of grains		Mean counts nucleoplasm	Mean counts cytoplasm
		Nucleoplasm	Cytoplasm		
Unirradiated cells	37	2016	1885	54.5	50.9
Nucleolus irradiated	50	1284	347	25.7	6.9
Whole nucleus irradiated	43	207	217	4.8	5.0
Nucleplasm irradiated	24	888	806	37.0	33.6

[a] From Harris (1970).
[b] Ultraviolet microbeam used as source of radiation.

order for expression of these genes to occur. Thus, Harris was led to suggest that nucleolar activity is essential for the transport of mRNA to the cytoplasm. This dependence was demonstrated by uv microbeam inactivation of the nucleolus which under these conditions will not transport. It is clear from autoradiographic experiments using [³H]uridine that in the absence of a functional nucleolus the RNA synthesized is not transported into the cytoplasm.

Therefore, the schedule of appearance of proteins is understandable. It follows that the nucleolus must have a transport role, which is also true of nonfused cells (Table 5.2) in which irradiation of the nucleolus with a microbeam inhibits transport of labeled RNA. These findings have led Harris to conclude that the nucleolus may have a more complex role than previously believed, which he has postulated to be a regulatory mechanism governing the flow of components of the ribosomal complex. It may be that the polydisperse RNA is altered by the nucleolus to form a functional messenger in the cytoplasm (Harris, 1976).

The migration of proteins during nuclear reactivation has been studied in detail by Ringertz and his collaborators (Carlsson *et al.*, 1974). Such investigations have shown that upon fusion of human HeLa cells with chick erythrocytes an enlargement and decondensation of the chick nucleus occurs. During the reactivating period, both histone and nonhistone proteins migrate from the human nucleus into the chick nucleus. The histone component is mainly composed of human H1 histone, whereas during this period there is a loss of the chick histone f2c.

V. CONCLUSIONS

Heterokaryons (in the context considered here) are multinucleate cells produced with the aid of fusogens such as Sendai virus and polyethylene glycol. They are transient structures and are ordinarily studied by methods of single cell analysis including histochemistry and autoradiography. The different nuclei can be identified morphologically or by the use of radioisotopes or fluorochromes.

Heterokaryons have been extremely valuable in analyzing membrane fluidity and have dramatically demonstrated the movement of cell surface components. They have also been of value in studying the regulation of the cell cycle through the use of cells in different phases of the cell cycle. Perhaps the most dramatic studies carried out in this system have been those in which the transfer of regulatory signals between nuclei have been studied. These investigations have established that when quiescent cells that do not synthesize RNA or DNA are fused with types that do, nuclear reactivation occurs. The exact mechanism responsible for this process remains to be clarified, but it appears to be the result of the migration of specific classes of histone proteins into the inactive nucleus. On the other hand, ''luxury'' functions are usually not suppressed in heterokaryons, and it is only with the subsequent breakdown of individual nuclei and formation of hybrids that they are regulated. A notable exception to this rule is the case of tyrosine aminotransferase, which is rapidly suppressed in hybrids. From these observations it must be concluded that regulation on several levels is occurring, with quite different consequences.

6

Chromosome Mapping

I. INTRODUCTION

The localization of mammalian genes to specific chromosomes and the establishment of linkage distances is a field that has proceeded at an extremely rapid pace in recent years. Although extensive chromosome mapping studies were carried out for many years on organisms utilized in classical genetics, such as corn and *Drosophila,* these approaches could not be readily adapted to mammals, especially humans. Thus, in humans, linkage studies consisted of a search for distorted segregation ratios in pedigrees of sufficient size and amounted to a rather limited enterprise (Siniscalco, 1979). The development of a number of innovative methods for localizing genes has completely revolutionized this speciality, even though traditional methods are still employed, combined with the new approaches. Although cell hybridization is the most widely used of these new methods, a number of other exciting techniques are presently under scrutiny. These include molecular hybridization of nucleic acid probes to metaphase chromosomes, DNA restriction analysis, and the use of transgenotes and microcells. Techniques applied at the *in vivo* level include mapping by the use of deletions and duplications (dosage effects) as a supplement to classical Mendelian analysis. Combined use of these techniques has enabled investigators to localize genes to every one of the 22 autosomes and 2 sex chromosomes of the human karyotype; the most detailed maps have collated data from *in vivo* and *in vitro* methods. The eventual goal of such studies would be a complete genetic blueprint of man with all regulatory and structural genes and all accessory material mapped so as to allow for a complete understanding and manipulation of the genetic material.

II. MAPPING BY PEDIGREE ANALYSIS

The classical method of gene mapping in man is to identify families in which alleles for at least two different characters are segregating and to identify parental and presumptive crossover types among the progeny. If the parental types exceed the nonparentals by a statistically significant fraction, there is evidence for genetic linkage. Usually data collected from such crosses are analyzed by the use of a probability ratio known as the "lod score," which is derived from the maximum likelihood method. This approach assigns a probability for the different linkage distances and yields a most likely distance between two loci (Siniscalco, 1979).

Pedigree analysis has a number of shortcomings. Because all-or-none characters are required, rare variants (inborn errors of metabolism, blood groups) are usually employed, and it is frequently difficult to obtain matings yielding enough data to make a judgment. This is further limited by the unavailability of information on coupling arrangements of the genes in the parents and by the inability to control matings. Although this approach may give information on linkage relationships, it cannot give information on chromosomal assignment, a fact that further limits its theoretical and practical utility. Thus, it is hardly surprising that in 1969 only seven linkage groups were identified in man (McKusick, 1969).

Traditional Mendelian analysis in man does allow for the assignment of genes to the X chromosome because of the particular father–daughter transmission of the X chromosome. Thus, even genes that fail to show linkage, such as X_g blood group and hemophilia, can be assigned to the same chromosome—a feat not possible, of course, in the case of the autosomes. Furthermore, indirect evidence for X linkage can be provided by a pattern of gene inactivation in heterozygotes resulting from the Lyon effect (see Chapters 2 and 13). This method was used to establish X linkage for the gene for the testicular feminization syndrome (Meyer *et al.*, 1975), in which the carrier female possesses two populations of cells: one with a full complement of testosterone receptors and another with no receptors.

Parenthetically, it should be mentioned that "Ohno's Rule" has provided additional evidence for X linkage in a number of cases. This is the observation (Ohno, 1973) that X linkage of genes is consistent for all mammals. Thus, determination of X linkage relationships in laboratory mammals (i.e., testicular feminization in the mouse) has served as strong evidence for assignment to the human X of this marker.

III. LINKAGE MAPPING USING SOMATIC CELL HYBRIDIZATION, CHROMOSOMAL VARIANTS, AND NUCLEIC ACID HYBRIDIZATION

A. Linkage Studies with Cell Hybrids

As long as 25 years ago there were suggestions that genes in man could be mapped using a "parasexual" analysis that had proved useful in fungal genetics (Pontecorvo, 1959). This technique, which was utilized in *Aspergillis,* takes advantage of the segregation of chromosomes following the fusion of two different strains. With the discovery of cell fusion techniques in the 1960s and the recognition that mouse–human hybrids preferentially segregated human chromosomes, the way was open to a full-scale exploitation of this method. During this period, improvements in chromosome staining added to the technical advances and in a short period a number of genes were localized to specific chromosomes.

The standard HAT selection method (Littlefield, 1964b; see Chapter 2) was used to obtain cellular hybrids. This approach has been supplemented by other selective methods (see Chapter 3). In addition to selective methods based on the elimination of both parents through the use of drug resistance factors, it is possible under some conditions to remove certain types (such as diploid fibroblasts) simply by virtue of their reduced plating efficiency. Thus, human fibroblasts hybridized with thioguanine-resistant mouse transformed cells can be selected using HAT medium followed by cloning of hybrid colonies (Davidson and Ephrussi, 1965).

Such somatic hybrids between different species are viable and both sets of information are expressed. In rodent–human hybrids, however, human chromosomes are lost and the result is that eventually a reduced segregant will be formed with most or all of the human chromosomes eliminated. Thus, the process simulates Mendelian segregation and assortment and represents an "alternative to sex." Three levels of gene localization may be carried out.

1. Synteny test—two gene loci are linked if they consistently assort together in a series of independently derived hybrid clones (Fig. 6.1). In such experiments many independent clones that retain particular human chromosomes are examined. By measuring the frequency with which pairs of genetic markers are lost or retained concordantly, it is possible to decide whether they are on the same chromosome, or "syntenic" (Fig. 6.2).

2. Assignment test—a gene is localized to a specific chromosome through its concordant segregation with that chromosome in a series of independently derived hybrids (Fig. 6.2).

3. Regional mapping—using a series of specific clones with chromosomal rearrangements, a particular gene can be localized to a restricted region of a chromosome.

By combining the somatic cell hybridization mapping with conventional pedigree studies, a composite map may be produced (for example, Fig. 6.3).

Among the most straightforward systems for chromosomal assignment is the use of selective markers. For instance, BUdR-resistant, thymidine kinase-deficient mouse cells, when hybridized with human fibroblasts, will proliferate in HAT medium. Continued growth in HAT will result in the retention of the human chromosome with the TK gene, whereas growth in BUdR will result in its loss through selection. Using this procedure, the TK gene was localized to

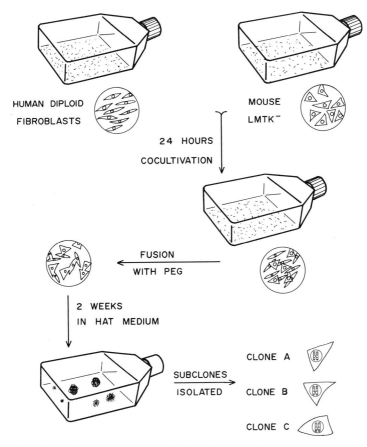

Fig. 6.1. Method for production of cell hybrids for use in chromosome mappings. Human fibroblasts are mixed together with mouse LMTK⁻ cells or with some other permanent rodent line carrying selectable genetic markers. Fusion is promoted through the use of polyethylene glycol, after which the cells are plated in HAT medium. Some of the cells undergo fusion forming first heterokaryons and then cell hybrids which proliferate in the selective HAT medium. The hybrid colonies that subsequently arise will undergo chromosome loss, retaining one or more human chromosomes. The distribution of human chromosomes is random so that individual hybrid clones will have dissimilar chromosomal content.

human chromosome 17 (McKusick and Ruddle, 1977). Several other genes, including HPRT (chromosome X) and APRT (chromosome 16), have been assigned in this fashion.

Another selective system that has been successfully exploited is the use of cytotoxic antibodies to cell surface antigens. Using polyvalent anti-human antiserum, Puck and colleagues (1971) demonstrated a linkage between lactate de-

HUMAN CHROMOSOMES

		1	2	3	4	5	6	7	8
	A	+	+	+	+	−	−	−	−
HYBRID CLONES	B	+	+	−	−	+	+	−	−
	C	+	−	+	−	+	−	+	−

Fig. 6.2. Determination of linkage relationships and chromosomal location of genes in hybrid cells. Gene locations are determined in this example by noting which gene products consistently associate with one another. In this example enzymes I and III are linked to the same chromosome (i.e., syntenic). In addition, they are both located on chromosome 2, whereas enzyme II is on chromosome number 1. (From Ruddle and Kucherlapati, 1974.)

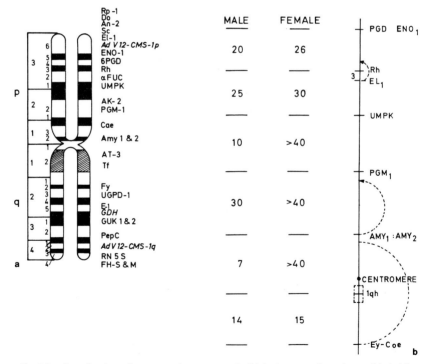

Fig. 6.3. Localization of genes on chromosome 1. (A) Assignments based on cell hybridization. (B) Map distances derived from pedigree analysis. Distances are greater in females than in males because of the greater incidence of crossing over. (From Vogel and Motulsky, 1980.)

hydrogenase (LDH) and a surface antigenic marker. More recently, using mono-clonal antibodies (see Chapter 9) produced from highly reduced mouse–human hybrids, Buck and Bodmer (1975) demonstrated the presence of a gene for a surface antigen on chromosome 11. In this approach, such cells are used as immunogens to elicit the production of antibodies against surface antigens coded by genes on the residual chromosomes present in the hybrid cells.

B. Use of Nucleic Acid Probes

If a radioactively labeled mRNA for a particular gene is available, the-oretically it should be possible to hybridize it to metaphase chromosome prepara-tions and thereby visualize the location of the gene in question. This approach would require the specific gene for gene association of the radioactive probe with the appropriate chromosomal regions. Although it is possible to localize the ribosomal RNA genes using this method (because of their redundancy), there are many technical and theoretical problems associated with the use of this method and results using it have been greeted with skepticism.

For instance, Price *et al.* (1972) claimed that they were able to localize the β, γ, and δ genes for hemoglobin to a B group chromosome and the α gene to chromosome 2. However, Bishop and Jones (1972) and Prensky and Holmquist (1973) have argued that the results cannot be valid. They assert that the limited number of copies of the genes in question could not possibly hybridize sufficient radioactive messenger to take up the amount of observed label in the time period involved in exposing the radiograms. They suggest that because of a bias in the selection of karyotypes an invalid result was obtained. According to their cal-culation, only genes with a very high degree of redundancy can hybridize suffi-cient messenger to be mapped by this approach.

The controversy concerning the localization of the hemoglobin loci has been resolved through improvements in the technology for the detection of gene sequences in cell hybrids. Gusella *et al.* (1979) have developed a hamster–human hybrid clone which retains only human chromosome 11, on which the β-globin gene was shown to reside. The original clone was treated with chromo-some breaking agents, and a panel of clones were selected that contained differ-ent combinations of the closely linked genetic markers lactate dehydrogenase type A, acid phosphatase 2, and the human antigenic surface markers, SA11-1, -2, and -3. Five clones were selected, each displaying a different terminal dele-tion, thus dividing chromosome 11 into six regions. To test for the presence of the globin gene sequence, DNA was prepared from the different clones by the Southern blot technique (Southern, 1975) and tested against hybridization probes from cloned β- and γ-globin genes on plasmids. The pattern of hybridization allowed a precise localization of the gene to a specific interband region of approximately 4500 kb (see Fig. 6.4). The authors commend these techniques to

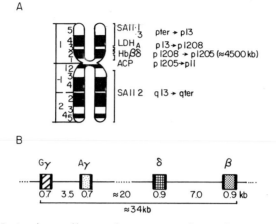

Fig. 6.4. (A) Regional map of human chromosome 11 showing location of β- and δ-globin structural genes. (B) Proposed gene map showing the location of the four globin genes with their respective introns and exons. (Adapted from Gusella *et al.,* 1979.)

fine structure resolution of the human genetic map. It would appear that their limitations are (1) the amount of effort involved in generating an appropriate panel of clones for the fine localization of a particular marker and (2) the obtaining of the gene in question in a cloned plasmid.

Localization of the immunoglobulin genes in both mouse and humans is a related problem that has occupied the energies of several research teams recently. This issue appears to be even more controversial than the mapping of the hemoglobin loci, in part due to the greater number of genes involved and the greater complexity of the system (see subsequent chapters).

Valbuena *et al.* (1978) tested DNA extracted from a panel of mouse–human hybrid clones lacking different mouse chromosomes in order to localize the genes for mouse immunoglobulin synthesis. Using cDNA probes for various light and heavy chains, they produced hybridization data that indicate that the C_H, C_λ, and C_κ genes reside within a group of several chromosomes. However, none of these is on chromosome 6 of the mouse, to which Swan *et al.* (1979) assign the immunoglobin κ light chain gene. These latter authors used a different approach—agarose gel electrophoresis, followed by hybridization with a nick-translated probe. They feel that this technique has a much higher resolution than that employed by Valbuena *et al.* (1978) and that by combining it with an appropriate panel of hybrid clones many different alleles should be localized.

D'Eustachio *et al.* (1980) used nucleic acid hybridization techniques to localize the genes for the constant regions of mouse heavy chains to chromosome 12. In these experiments DNA from mouse–hampster hybrids was extracted from cells, digested with restriction endonucleases, fractionated by agarose gel electrophoresis, and bound to nitrocellulose filters. When the filters were tested against DNA probes for the λ_{2b}, μ, and α heavy chain regions, it was found that only hybrids retaining mouse chromosome 12—in fact, the distal half—reacted with the probes. These results are consistent with a previous finding (Meo *et al.*, 1980) from linkage studies with mice carrying a translocation in band 12B1. Because such experiments directly test for the presence of the gene, they avoid ambiguities posed by approaches that measure the gene product (i.e., immunoglobulin or immunoglobulin messenger).

Determination of the chromosomal localization of human immunoglobin genes has also resulted in some controversy. Smith and Hirschhorn (1978) pinpoint the gene for human IgG heavy chains to human chromosome 6, but their results conflict with those of Croce *et al.* (1979a), who assigned the same gene to chromosome 14. Because Smith and Hirschhorn used mouse fibroblast–human lymphoblast hybrids, which ordinarily are suppressed for immunoglobulin synthesis (Chapter 9), it is difficult to see precisely how such hybrids could allow for the localization of these alleles. Croce *et al.* (1979a), on the other hand, used mouse myeloma by human lymphoblast hybrids that produce antibody chains specific to both parents and are therefore more appropriate experimental material. Both groups of workers measured immunoglobulin chain production rather than the presence of the gene through the use of hybridization probes.

Although the use of nucleic acid probes to localize genes is faced with a number of limitations (such as the technical problems of mapping genes present in single copies), several new developments have resulted in significant improvements in this approach. Ruddle (1981) has reviewed recent innovations, including (1) the use of ^{125}I-labeled nucleotide, which when introduced into the DNA produces radioactive probes of very high specific activity; (2) single-stranded DNA probes that do not hybridize to themselves; (3) fluorescent antibodies to probes that have been altered so as to make them better antigens; and (4) the use of dextran sulfate in the hybridization mixture to increase the efficiency of hybridization.

C. The Lepore Method

If two genes are contiguous (or closely linked) and a mutation resulting from a deletion of the midregion joining the two genes occurs, a hybrid protein will result and will contain a portion of both molecules (Baglioni, 1962). This is evidence of linkage and can be used to determine the location of the genes with respect to one another. This has occurred in the case of the Lepore hemoglobin and will be discussed in more detail later (Fig. 6.5).

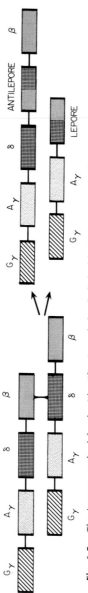

Fig. 6.5. The Lepore method for the identification of closely linked loci. An instance of nonhomologous pairing results in an unequal crossing over, which causes the formation of an abnormal gene product (in this case a hemoglobin with fused β and δ chains). (From McKusick and Ruddle, 1977.)

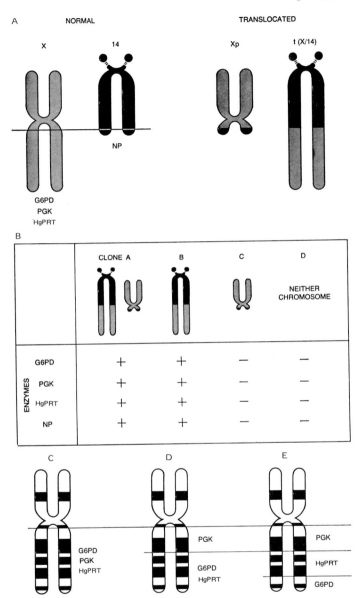

Fig. 6.6. Use of translocations in the mapping of genes through the use of somatic cell hybridization. (A) An X-14 translocation. (B) The four types of clones obtained with respect to the presence or absence of the translocated chromosomes and the resultant phenotypes of four human enzymes analyzed in these crosses. These data demonstrate that G6PD, PGK, and HPRT are all on the X chromosome and NP is on chromosome 14. (C, D, and E) Data obtained from a composite of experiments involving different translocations that establish that the order of the genes is as shown in E (Ruddle and Kucherlapati, 1974).

D. Gene Dosage Methods

This can be performed using either deletions [Rh and phosphogluconate dehydrogenase (6PGD) were shown to be on the same chromosome by examination of a patient with a partial deletion of chromosome 1 in some of his red cells] or duplications (superoxide dismutase was present in 50% greater quantities in cells from Down's syndrome patients and is known to be on chromosome 21).

E. Mapping through Induction of Chromosome Breaks in Cell Hybrids

Chromosome breakage can be induced *in vitro;* this has been done in a manner analogous to that described earlier for the hemoglobin loci. Using adenovirus 12, which breaks specifically chromosome 17, investigators have been able to localize the thymidine kinase gene. These experiments used a series of hybrids lacking different regions of chromosome 17; analysis of the data involved correlation of the presence of the phenotype with the presence of a particular band (Fig. 6.6; Ruddle and Kucherlapati, 1974).

F. Use of Cellular Fragments

The use of isolated metaphase chromosomes as a technique for localizing genetic information was originated by McBride and Ozer (1973) and has already been considered (Chapter 4). By adding preparations of partially purified chromosomes under appropriate selective conditions, it is possible to obtain approximately 1 in 5 million transformants. Although it is not clear at this time whether the information is inserted in the host genome, such transgenotes may be stable or unstable and have *never* been observed to contain an intact chromosome or recognizable fragment from the donor line.

More recent observations (McKusick, 1980) indicate that stability of the transferred chromosomal element is acquired by integration into a host chromosomal segment bearing a host centromere and that the point of integration is at the donor centromere, which is lost in the process.

IV. CONCLUSIONS

We still do not have sufficient data on the properties of the human map to draw many sweeping theoretical conclusions. However, one interesting point has been discussed in detail by Vogel and Motulsky (1980): the existence of gene clusters possessing related functions. This problem has a long history derived from investigations on microbial systems as well as in eukaryotes. The Jacob–Monod

(1961) model of regulation in bacteria presented a theoretical basis for close linkage of related gene loci in order to bring about an economical coordinate regulation. It was subsequently shown that in bacteria many metabolic pathways involving long series of reactions are under coordinate control (i.e., histidine biosynthesis) resulting from close linkage of the genes responsible for the enzymes involved. These genes are regulated by a single operator and a single regulator gene.

If extensive clustering of related functions occurred in the human genetic map, this would have important implications for gene regulation. Although at first glance the distribution of genes in man appears to be random, there are some important examples of gene clusters. These include γ-, δ-, and β-hemoglobin, immunoglobulin loci, the major histocompatibility cluster, and the genes for protan and deutan color blindness. However, these observations have not enabled us to formulate a model for the regulation of the synthesis of products such as hemoglobin, and the fact that the α locus is unlinked to the β–γ–δ complex suggests that other factors may be involved. A more parsimonious explanation for this clustering may be an evolutionary explanation: because these loci share many amino acid sequences in common and are most certainly derived from a common evolutionary precursor, it appears more reasonable to postulate a common origin through gene duplication.

The present direct application of this information to human illness is restricted to genetic counseling. Table 6.1 and Fig. 6.7 provide a summary of recent data from the human chromosome map. One limited use of the present technology is diagnosis *in utero* of genetic disorders that do not display a detectable phenotype in amniotic cells. For instance, in the case of a gene closely linked to the β-hemoglobin locus, it would be possible to give assistance to a couple, both of whom were heterozygous for sickle cell anemia (Hb^a/Hb^s), if the prospective mother were also heterozygous for a closely linked gene that could be detected in the fibroblasts and if the coupling relationship in the mother were known. This method of counseling puts so many specific requirements on the particular situation that its utility is at present quite limited. It has been employed in prenatal diagnosis by taking advantage of the close linkage of hemophilia A and G6PD (McCurdy, 1971) and of secretor and myotonic muscular dystrophy (Insley *et al.*, 1976). However, a more direct molecular approach may make this technique obsolete. It has been shown that a modification of the DNA restriction pattern occurs in the case of the sickle cell β-globin gene, enabling a prenatal diagnosis of *Hpa*I-digested DNA from amniotic fluid cells. Thus, Kan and Dozy (1978) have successfully diagnosed the fetus' Hb^s globin genotype in several pregnancies at risk. This change in the *Hpa*I restriction pattern (to a 13-kb fragment) appears to be due to a polymorphism associated with the Hb^s gene in a state of linkage disequilibrium. Linkage disequilibrium is a condition that arises when two closely linked mutations occur preferentially on the same homolog followed

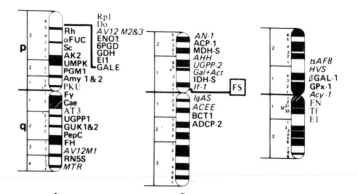

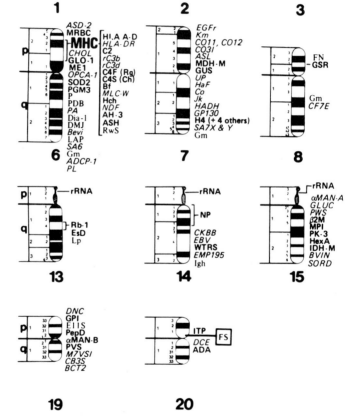

Fig. 6.7. A diagrammatic synopsis of the gene map of the human chromosomes. (From McKusick, 1980). The banding patterns and numbering of banded regions are those given in the International System for Human Cytogenetics Nomenclature 1978. An assignment is considered confirmed if found in two laboratories or several families; it is considered provisional if based on evidence from only one laboratory. Inconsistent assignments based on conflicting evidence and assignments for which the evidence is weaker than that for provisional assignment are separately indicated (also termed "tentative" or "in limbo"). In Table 6.1 following the figure, the chromosomal and in some instances the regional location of the gene is given.

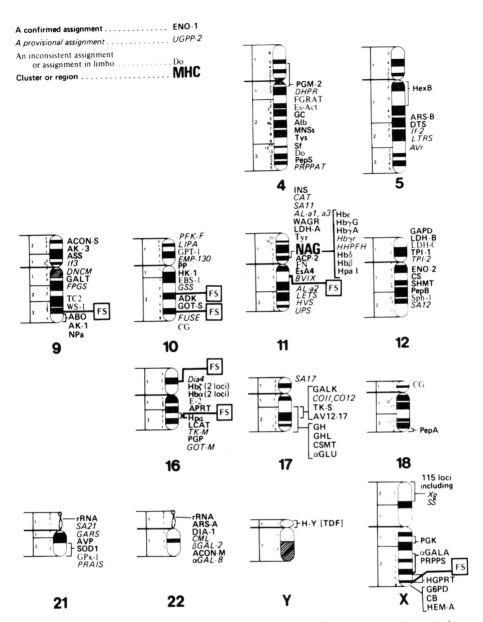

A confirmed assignment . . . . ENO·1

A provisional assignment . . . . UGPP·2

An inconsistent assignment
   or assignment in limbo . . . . Do

Cluster or region . . . . **MHC**

Those chromosomal or regional locations indicated in parentheses represent assignments by indirect means, e.g., linkage to a locus directly assigned. Also indicated are the methods of assignment: F, family linkage studies; S, somatic cell hybrid studies; A, *in situ* annealing ("hybridization"); H-S, molecular hybridization in solution; RE, restriction endonuclease mapping; D, deletion mapping or dosage effects; AAS, deduction from amino acid sequence (Lepore phenomenon); LD, linkage disequilibrium; V, virus effects; Ch, chromosome change; OT, ovarian teratoma; EM, exclusion mapping; H, homology; R, radiation-induced gene segregation. For the most recent revision of this information, see McKusick (1982).

TABLE 6.1 Gene Locus Symbols Used in Fig. 6.7[a]

Symbol	Description
ABO	ABO blood group—9q34 (F)
ACEE	Acetylcholinesterase expression—chr. 2 (D)
ACON-M	Aconitase, mitochondrial—22q11-qter (S)
ACON-S	Aconitase, soluble—9pter-p13 (S)
ACP1	Acid phosphatase-1—2p23 (D,S)
ACP2	Acid phosphatase-2—11p12-cen (S)
ACY1	Aminoacylase-1—3pter-q13 (S)
ADA	Adenosine deaminase—20q13-qter (S,D)
ADCP1	Adenosine deaminase complexing protein-1—chr. 6 (S)
ADCP2	Adenosine deaminase complexing protein-2—chr. 2 (S)
ADK	Adenosine kinase—10q11-q24 (S,D-EM)
AHH	Aryl hydrocarbon hydroxylase—2p (S)
AH3	Adrenal hyperplasia III (21-hydroxylase deficiency) (6p2105-6p23) (F)
AK1	Adenylate kinase-1 (soluble)—9q34 (F,S,D)
AK2	Adenylate kinase-2 (motochondrial)—1p3-p34 (S,F,R)
AK3	Adenylate kinase-3 (mitochondrial)—9pter-p13 (S)
AL	Lethal antigen: 3 loci—a1, a3 on 11p13-pter; a2 on 11q13-qter (S)
Alb	Albumin—4q11-q13 (F)
AMY1	Amylase, salivary—1p (F-F)
AMY2	Amylase, pancreatic—1p1 (linked to ACP1) (F)
An1	Aniridia, type 1 (chr. 2: linked to ACP1) (F)
APRT	Adenine phosphoribosyltransferase—16q (S,D)
ARS-A	Arylsulfatase A—chr. 22 (S)
ARS-B	Arylsulfatase B—chr. 5 (S)
ASD2	Atrial septal defect, secundum type (chr. 6: linked to HLA) (F)
ASH	Asymmetric septal hypertrophy (chr. 6: linked to HLA) (F)
ASL	Argininosuccinate lyase—7pter-q22 (S)
ASS	Argininosuccinate synthetase—chr. 9 (S,D)
AT3	Antithrombin III (chr. 1) (F)
AV12M1	Adenovirus-12 chromosome modification site-1—1q42-43 (V)
AV12M2	Adenovirus-12 chromosome modification site-2—1p36 (V)
AV12M3	Adenovirus-12 chromosome modification site-3—1q21 (V)
AV12-17	Adenovirus-12 chromosome modification site-17—17q21-q22 (V)
AVP	Antiviral protein—21q21-qter (S,D)
AVr	Antiviral state regulator—chr. 5 (D)
β2M (B2M)	Beta-2-microglobulin—15q14-q21 (S)
BCT-1	Branched chain amino acid transferase-1—chr. 12 (S)
BCT-2	Branched chain amino acid transferase-2—chr. 19 (S)
BEV1	Baboon M7 virus infection—chr. 6 (S)
BF	Properdin factor B—chr. 6 (in MHC) (F)
BVIN	BALB virus induction, N-tropic—chr. 15 (S)
BVIX	BALB virus induction, xenotropic—chr. 11 (S)
C2	Complement component-2—chr. 6 (in MHC) (F)
C4F	Complement component-4 fast—chr. 6 (in MHC) (F)
C4S	Complement component-4 slow—chr. 6 (in MHC) (F)
Cae	Cataract, zonular pulverulent (chr. 1: linked to Fy) (F)
CAT	Catalase—11p (S)
CB	Colorblindness (deutan and protan) (Xq28) (F)
CB3S	Coxsackie B3 virus susceptibility—chr. 19 (S)
CF7E	Clotting factor VII expression (chr. 8) (D)
CG	Chorionic gonadotropin (chr. 10 and 18: chr. 5 or 6) (S,REb)
Ch	Chido blood group—same as C4S (F)
CHOL	Hereditary hypercholesterolemia—chr. 6 (?linked to HLA) (F)
CKBB	Creatine kinase, brain type—chr. 14 (S)
CML	Chronic myeloid leukemia—22q12 (Ch)
Co	Colton blood group (chr. 7) (D,F)
COI1	Collagen I alpha-1 chain—chr. 7 and 17 (S,M)
COI2	Collagen I alpha-2 chain—chr. 7 and 17 (S,M)
COI31	Collagen III alpha-1 chain—chr. 7 (S)
CS	Citrate synthase, mitochondrial—chr. 12 (S)
CSMT (or CSH)	Chorionic somatomammotropin—(chr. 17) (S)
DCE	Desmosterol-to-cholesterol enzyme—chr. 20 (F)
DHPR	Quinoid hydropteridine reductase—chr. 4 (S)
Dia-1	NADH-diaphorase—chr. 22 (S)
DIA-4	Diaphorase-4—chr. 16 (S)
DMJ	Juvenile diabetes mellitus (chr. 6: ?linked to HLA) (F,LD)
DNC	Lysosomal DNA-ase—chr. 19 (S)
DNCM	Cytoplasmic membrane DNA—9qh (H-A)
Do	Dombrock blood group (?chr. 1 or 4) (F)
DTS	Diphtheria toxin sensitivity—5q15-qter (S)
E1	Pseudocholinesterase-1 (?chr. 3: linked to Tf) (F)
E2	Pseudocholinesterase-2—16cen-q22 (F)
E11S	Echo 11 sensitivity—19q (S)
EBS1	Epidermolysis bullosa, Ogna type (chr. 10) (F)
EBV	Epstein-Barr virus integration site—chr. 14 (S)
EGFR	Epidermal growth factor, receptor for—chr. 7 (S)
E11	Elliptocytosis-1—(1p; linked to Rh) (F)
EMPI30	External membrane protein-130—chr. 10 (S)
EMPI95	External membrane protein-195—chr. 14 (S)
ENO1	Enolase-1—1p36-1pter (S,F,R)
ENO2	Enolase-2—chr. 12 (S)
Es-Act	Esterase activator—chr. 4 or 5 (S)
EsA4	Esterase-A4—11cen-q22 (S)
EsD	Esterase D—13q14 (S,F,D)
FGRAT	Formylglycinamide ribotide amidotransferase—chr. 4 or 5 (S)
FH	Fumarate hydratase—1q42-qter (S,R)
FN	Fibronectin—chr. 8, 11 (S)
FPGS	Folypolyglutamate synthetase—chr. 9 (S)
FS	Fragile site, observed in cultured cells, with or without folate deficient medium, or BrdU—2q11: 10q23; 10q25; 11q13; 16p124; 16q22; 20p11; Xq27 (F,S)
αFUC (FUCA)	Alpha-L-fucosidase—1p32-p34 (S,F,R)
FUSE	Polykaryocytosis inducer—chr. 10 (S)
Fy	Duffy blood group—1q13 (F,Fc)
Gal+Act	Galactose + activator—chr. 2 (S)
αGALA	Alpha-galactosidase A (Fabry disease)—Xq22-q24 (F)
αGALB	Alpha-galactosidase B—22q13-qter (S)
βGAL-1	Beta-galactosidase-1—3pter-q13 (S)
βGAL-2	Beta-galactosidase-2—22q13-qter (S)
GALE	Galactose-4-epimerase—1p21-pter (S,LD)
GALK	Galactokinase—17q21-q22 (S,C,R)
GALT	Galactose-1-phosphate uridyltransferase—9p13 or 9p22 (S)
GAPD	Glyceraldehyde-3-phosphate dehydrogenase—12p122-pter (S,D)
GARS	Glycinamide ribonucleotide synthetase—chr. 21 (S)
GC	Group-specific component—4q11-q13 (F,Fc)
GDH	Glucose dehydrogenase—1p21-pter (1p32-pter) (S)
GH	Growth hormone—chr. 17 (S)
GHL	Growth hormone like—chr. 17 (S,RE)
αGLU (GLUA)	Alpha-glucosidase—chr. 17 (S)
GLUC	Neutral alpha-glucosidase C—chr. 15 (S)
GLO1	Glyoxalase I—6p21-6p22 (F,S)
GOT-M	Glutamate oxaloacetate transaminase, mitochondrial—chr. 16 (S)
GOT-S	Glutamate oxaloacetate transaminase, soluble—10q24-q26 (S,D)
G6PD	Glucose-6-phosphate dehydrogenase—Xq28 (F,S,R)
GPI30	Granulocyte glycoprotein—7q22-qter (D)
GPI	Glucosephosphate isomerase—chr. 19 (S,D)
GPT1	Glutamate pyruvate transaminase, soluble—chr. 10 (F)
GPx1	Glutathione peroxidase-1—3p13-q12 (S)
GSR	Glutathione reductase—8p21 (S,D)
GSS	Glutamate-gamma-semialdehyde synthetase—chr. 10 (S)
Gm	Immunoglobulin heavy chain—chr. 6, 7, 8 (see Igh) (S,Fc,S)
GUK1 & 2	Guanylate kinase-1 & 2—1q32-q42 (S)
GUS	Beta-glucuronidase—chr. 7 (S)
H4	Histone H4 and 4 other histone genes—chr. 7 (A)
HADH	Hydroxyacyl-CoA dehydrogenase—chr. 7 (S)
HaF	Hageman factor—7q35 (D)
Hbα(HBA)	Hemoglobin alpha chain—chr. 16 (S-HS)
Hbβ(HBB)	Hemoglobin beta chain—11p205-p1208 (LD,AAS,F,RE,S)
Hbδ(HBD)	Hemoglobin delta chain—11p1205-p1208 (LD,AAS,F,RE,S)

Symbol	Description
Hbγr (HBG)	Hemoglobin gamma regulator—11p1205-p1208 (RE)
Hbγ(HBG)	Hemoglobin gamma chains—11p1205-p1208 (AAS,RE)
Hbε(HBE)	Hemoglobin epsilon chain—11p1205-p1208 (AAS,RE)
HBζ(HBZ)	Hemoglobin zeta chain—chr. 16 (RE)
Hch	Hemochromatosis (chr. 6; linked to HLA) (LD,F)
HEM-A	Classic hemophilia—Xq28 (F)
HexA	Hexosaminidase A—15q22-15qter (S)
HexB	Hexosaminidase B—5cen-q13 (S)
HGPRT	Hypoxanthine-guanine phosphoribosyltransferase—Xq26-qter (F,S,R)
HHPFH	Heterocellular hereditary persistence of fetal hemoglobin—11p1205-p1208 (F)
HK1	Hexokinase-1—10pter-q24 (F)
HLA(A-D)	Human leukocyte antigens—6p2105-p23 (F)
HLA-DR	Human leukocyte antigen, D-related—6p2105-p23 (F)
Hpα	Haptoglobin, alpha—16 (Fc)
Hpa1	Hpa 1 restriction endonuclease polymorphism—11p1205-p1208 (RE)
HVS	Herpes virus sensitivity (chr. 3 and 11) (S)
H-Y	Y histocompatibility antigen (Y chr.) (F)
IDH-M	Isocitrate dehydrogenase, mitochondrial—15q21-qter (S)
IDH-S	Isocitrate dehydrogenase, soluble—2q11 or 2q32-qter (S)
If1	Interferon-1—2p23-qter (S)
If2	Interferon-2 (chr. 5) (S)
If3	Interferon-3 (chr. 9) (S)
IgAS	Immunoglobulin heavy chains attachment site—chr. 2 (S)
Igh	Immunoglobulin heavy chains (mu, gamma, alpha)—chr. 14 (see Gm) (S)
Ins	Insulin—11p (S)
ITP	Inosine triphosphatase—20p (S)
Jk	Kidd blood group—7q (Fc)
Km	Kappa immunoglobulin light chains, Inv (chr. 7) (F,Fc)
LAP	Laryngeal adductor paralysis—(chr. 6; linked to HLA) (F)
LCAT	Lecithin-cholesterol acyltransferase—(16q22; linked to Hp alpha) (F,LD)
LDH-A	Lactate dehydrogenase A—11p1203-p1208 (S)
LDH-B	Lactate dehydrogenase B—12p121-p122 (S,D)
LDH-C	Lactate dehydrogenase C—(12p; linked to LDH-B in pigeon) (H)
LIPA	Lysosomal acid lipase-A—chr. 10 (S)
Lp	Lipoprotein-Lp—chr. 13 (F)
LTRS	Leucyl-tRNA synthetase—chr. 5 (S)
β2M (B2M)	Beta-2-microglobulin—15q22-15qter (15q12-q21) (S)
M7VS1	Baboon M7 virus sensitivity-1—chr. 19 (S)
αMAN-A	Cytoplasmic alpha-D-mannosidase—15q11-qter (S)
αMAN-B	Lysosomal alpha-D-mannosidase—19pter-q13 (S)
MDH-M	Malate dehydrogenase, mitochondrial—7p22-q22 (S)
MDH-S	Malate dehydrogenase, soluble—2p23 (S)
ME1	Malic enzyme, soluble—6p21-q16 (S)
MHC	Major histocompatibility complex—6p2105-p23 (F,S)
MLC-W	Mixed lymphocyte culture, weak (chr. 6) (F)
MNSs	MNSs blood group—4q (F,Fc)
MPI	Mannosephosphate isomerase—15q22-qter (S)
MRBC	Monkey red blood cell receptor—chr. 6 (S)
MTR	5-Methyltetrahydrofolate: L-homocysteine S-methyltransferase, or tetrahydropteroyl-glutamate methyltransferase—chr. 1 (S)
NAG	Non-alpha globin region—12p1205-1208 (S,RE)
NDF	Neutrophil differentiation factor (chr. 6) (LD)
NP	Nucleoside phosphorylase—14q13 (S,D)
NPa	Nail-patella syndrome—(9q3; linked to ABO) (F)
OPCA1	Olivopontocerebellar atrophy 1—(chr. 6; linked to HLA) (F)
P	P blood group (chr. 6) (S,F)
PA	Pasminogen activator—(chr. 6) (S)
PDB	Paget disease of bone—(chr. 6; ?linked to HLA) (F)
PepA	Peptidase A—18q23-18qter (S,D)
PepB	Peptidase B—12q21 (S)
PepC	Peptidase C—1q25, or 1q42 (S,R)
PepD	Peptidase D—(chr. 19) (S)
PepS	Peptidase S—4pter-q12 (S)
6PGD	6-phosphogluconate dehydrogenase—1p34-pter (F-S)
PGK	Phosphoglycerate kinase—Xq13 (F,S)
PGM1	Phosphoglucomutase-1—1p32; 1p221-p311; 1p33-p34 (F,S,R)
PGM2	Phosphoglucomutase-2—4p14-q12 (S)
PGM3	Phosphoglucomutase-3—6q (S,F,OT)
PGP	Phosphoglycolate phosphatase—16p (S)
PK3	Pyruvate kinase-3—15q14-qter (S)
PKU	Phenylketonuria (1p; linked to AMY) (F)
PL	Prolactin—chr. 6 (S)
PP	Inorganic pyrophosphatase—10pter-q24 (S)
PRPPAT	Phosphoribosylpyrophosphate amidotransferase—4pter-q21 (S)
PRPPS	Phosphoribosylpyrophosphate synthetase—X chr. (S)
PRAIS	Phosphoribosylaminoimidazole synthetase—chr. 21 (S)
PVS	Polio virus sensitivity—19q (S)
PWS	Prader-Willi syndrome—15q11-q12 (Ch)
RB1	Retinoblastoma-1—13q12-q14; 13q21-22 (Ch)
rC3b	Receptor for C3b—chr. 6 (in MHC) (S)
rC3d	Receptor for C3d—chr. 6 (in MHC) (S)
Rg	Rodgers blood group—same as C4F (F)
Rh	Rhesus blood group (1p32-pter) (F,S,D)
RN5S	5S RNA gene(s)—1q42-q43 (A)
RP1	Retinitis pigmentosa-1 (chr. 1) (F)
rRNA	Ribosomal RNA—13p12, 14p12, 15p12, 21p12, 22p12 (A)
RwS	Ragweed sensitivity—(chr. 6; ?linked to HLA) (F)
SA6	Surface antigen 6—chr. 6 (S)
SA7	Surface antigen 7—7p12-pter (S)
SA11	Surface antigen 11—11p (S)
SA12	Surface antigen 12—chr. 12 (S)
SA17	Surface antigen 17—chr. 17 (S)
SA21	Surface antigen 21—chr. 21 (S)
Sc	Scianna blood group—(1p32-p34) (F)
Sf	Stoltzfus blood group—(4q; linked to MNSs) (F)
SHMT	Serine hydroxymethyltransferase—chr. 12 (S)
SOD1	Superoxide dismutase, soluble—21q211 (S,D)
SOD2	Superoxide dismutase, mitochondrial—6q21 (S)
SORD	Sorbitol dehydrogenase—15pter-q21 (S)
Sph1	Spherocytosis, Denver type (8p11 or chr. 12) (Fc)
SS	Steroid sulfatase (?Xp22-pter) (F,S)
TC2	Transcobalamin II—9q (?linked to ABO) (F)
TDF	Testis determining factor—prob. same as H-Y (F)
Tf	Transferrin—?chr. 3 (H)
TK-M	Thymidine kinase, mitochondrial—chr. 16 (S)
TK-S	Thymidine kinase, soluble—17q21-q22 (S,C,R)
TPI-1 & 2	Triosephosphate isomerase-1 & 2—TPI-1 on 12p12.2-pter (S)
tsAF8	Temperature-sensitive (AF8) complement—chr. 3 (S)
Tyr	Tyrosinase—(?11p) (H)
Tys	Sclerotylosis—(4q; linked to MNSs) (F)
UGPP1	Uridyl diphosphate glucose pyrophosphorylase-1—1q21-q23 (S,R)
UGPP2	Uridyl diphosphate glucose pyrophosphorylase-2—chr. 2 (S)
UMPK	Uridine monophosphate kinase—1p32 (S,R)
UP	Uridine phosphorylase—chr. 7 (S)
UPS	Uroporphyrinogen I synthase—chr. 11 (S)
WAGR	Wilms tumor–aniridia/ambiguous genitalia/mental retardation—11p13 (Ch)
WTRS	Tryptophanyl-tRNA synthetase—chr. 14 (S)
WS1	Waardenburg syndrome-1—(chr. 9; ?linked to ABO) (F)
Xg	Xg blood group (X chr.; ?Xp2) (F)

by an inadequate number of generations required to randomize the alleles through crossing over. Ruddle (1981) believes that such associations may turn out to be frequent and quite valuable in prenatal molecular genetic diagnosis and in the mapping of many human genetic disorders.

This brings us back to the original question of the justification of these studies. Although their immediate applicability to basic or practical scientific questions appears to be rather limited, this situation may change in the near future. Unquestionably, the surface of the human genome map has only been scratched. It is reasonable to assume that there are approximately 50,000 structural genes in humans. As more and more of these are mapped, much of the mechanism of regulation and differentiation will doubtless become apparent (McKusick, 1980). As the use of recombinant DNA and restriction mapping in mammalian molecular genetics becomes more refined, a number of projects will be made feasible. The insertion of particular genetic sequences will require selective systems, and the availability of closely linked markers (antigens, biosynthesis enzymes, membrane permeability factors) will greatly facilitate such studies. The consequences of direct intervention to alter the cell's genetic constitution are interesting to contemplate.

7

Differentiation in Cultured Cells: Liver Cells

I. INTRODUCTION

The control of gene expression in development is one of modern biology's most central issues. It is currently the subject of a substantial effort because the unraveling of this problem would have tremendous theoretical implications and, on a purely practical level, because aging, cancer, immune disorders, and many deleterious genetic conditions have their basis in changes related to the control of differentiation.

A voluminous body of information on descriptive embryology has accumulated over the course of the last century and has served as a framework for cellular and molecular investigations. These investigations appear to be the most promising means for resolving the actual mechanisms of differentiation. Cell culture has figured prominently among these approaches even though there are some questions that it will not answer, such as the developmental basis of complex structures and the expression of patterns requiring a chronological sequence of events. For example, the size and shape of a species must be regulated by genes that can dictate within very tight limits the three-dimensional structure of organ systems and of all bodily components. Because this three-dimensional relationship is destroyed in tissue culture systems, it is impossible to answer questions regarding the nature of its regulation. Another example that is inaccessible to tissue culture methodology is the intricate pattern of events that occur in an ordered framework in embryonic development. Organ systems are laid down and often remodeled and tissues are assembled on top of previously created structures, all in perfect chronological order. Because of the disruption caused by tissue culture procedures, it appears to be impossible to gain an understanding of these time-related processes through cell culture techniques.

The basic question of differentiation that can be studied using somatic cell genetics is, How do structural genes in different tissues regulate the intermittent expression of gene product? Whatever the exact mechanism, it certainly does not involve the loss of information, as shown by the nuclear transplantation studies of Briggs and King (1952) and Gurdon (1974). The authors developed techniques by which amphibian eggs could be successfully enucleated and implanted with nuclei from different stages of development. Given appropriate conditions, normal fertile clones of animals can be obtained from nuclei taken from blastula stage embryos. In a long series of experiments it was demonstrated that nuclei from progressively later stages of differentiation were less and less able to initiate normal development when transplanted into enucleated eggs. However, nuclei from the brush border cells of the larval gut can (in the case of *Xenopus*) bring about completely normal development. Although the percentage of successful takes is extremely small, these nuclei are derived from a highly differentiated cell type, and gene regulation must therefore proceed by some pattern of control in which structural genes are acted upon by other genetic elements while retaining their integrity within the genome.

The development of the operon model of bacterial gene regulation (Jacob and Monod, 1961) is one of the intellectual triumphs of the twentieth century and has exerted much influence on the thinking of workers in the area of eukaryotic control. Although some evidence exists for operator mutants in lower eukaryotes such as in yeast (Bollon, 1980), there are no data available at present to support the existence of an operator-type system in mammalian cells. This is due in part to the paucity of long inducible metabolic sequences that would serve as likely candidates for coordinate control of a group of metabolically related genes. For example, there is no histidine operon in mammalian cells because mammals have an absolute requirement for histidine and lack most of the enzymes necessary for its synthesis. There is, however, evidence from cell hybridization suggesting the existence of diffusible repressors by which differentiated characters can be regulated. The discussion of this system and its alternatives will be the main topic of this chapter.

Theoretically, there are three mechanisms by which structural genes could be controlled in the course of differentiation (as reviewed by Davis and Adelberg, 1973):

1. Autonomous expression of structural genes—the expression of the gene is not dependent on any outside entity once it is activated (for instance, by removal of a specifically chromosomally bound inhibitor).

2. Continuous production of an activator—in this case, *another* gene is the primary activation site causing production of a specific, diffusible molecule. In order for the structural gene to produce its product, the activator molecule must be constantly produced. If the activator gene is deleted, or ceases to function, the particular differentiated function will be extinguished.

3. Discontinuous production of a repressor—in this model, nuclear structural genes are activated in the course of development by the removal of specific repressor molecules.

Through the use of cell hybrids these three hypotheses are experimentally distinguishable. However, many problems must be recognized when this approach is utilized. Among these are

1. The question of the relevance of interspecific hybrids to our understanding of differentiation.

2. The inability of cell hybrids to produce a differentiated gene product because of an absence of ancillary features rather than because of gene inactivation (i.e., immunoglobulin-producing hybrid cells might lack the appropriate secretory apparatus even though possessing the structural genes).

3. The masking of surface properties by secondary factors unrelated to differentiation (for instance, hybrids between cells that grow in suspension and cells that grow attached).

II. LIVER CELLS AND THEIR HYBRIDS IN THE STUDY OF DIFFERENTIATION

Although a variety of tissue types have been shown to express differentiated functions *in vivo,* one of the most widely utilized in this regard is the liver. This includes both normal hepatocytes and tumor cells that were initially derived from the liver and retain at least some of their tissue-specific capacities (Fig. 7.1).

Perhaps one reason for the great interest in this tissue is because it possesses a large number of enzymes, many of which can be easily measured by standard biochemical procedures. In addition, selective systems have been designed using liver-specific markers, which take advantage of the liver's capacity to synthesize amino acids or to metabolize toxic compounds (Hankinson, 1979; Choo and Cotton, 1979). Table 7.1 summarizes some of the studies that have been carried out on hepatic cells in culture.

A. Tyrosine Aminotransferase

This liver-specific enzyme (TAT; EC 2.6.1.5) has been studied by many investigators, mainly in hepatoma tissue culture (HTC) cells (an *in vitro*-adapted hepatoma produced with chemical carcinogens in a buffalo rat). Ordinarily, TAT maintains a low basal level in liver cells or transformed derivatives. However, in the presence of glucocorticosteroids there is a substantial increase in enzyme activity to approximately 10- to 15-fold the previous level. Both the presence of the enzyme and its inducibility are liver-specific characters. The induction pro-

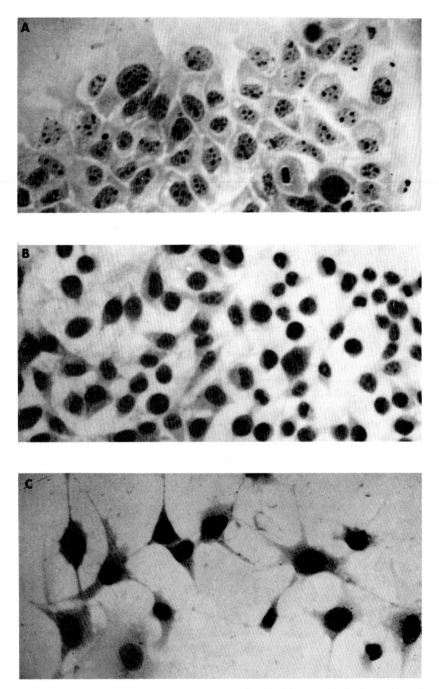

Fig. 7.1. Morphological appearance of liver cells, fibroblasts, and hybrids in culture (Rintoul *et al.*, 1973b). (A) Mouse fetal liver cells; (B) LMTK⁻ fibroblasts; (C) FL hybrid.

cess has been studied using inhibitors of RNA synthesis (Steinberg *et al.*, 1975) and the conclusion has been reached that glucocorticosteroids act by modulation of TAT messenger RNA levels (Tomkins *et al.*, 1974). Actinomycin D causes a phenomenon referred to as "superinduction" when the antibiotic is applied to the cell following glucocorticosteroid-mediated induction. The Tomkins model suggests that actinomycin D blocks the synthesis of labile RNA that is involved in prevention of the translation of TAT messenger. This results in an increase in the level of the enzyme above the induced level, because the mRNA for enzyme is assumed to be more stable. Steroid inducers (such as dexamethasone) block the action of the repressor. In high doses, actinomycin D blocks the transcription of repressor molecules that are rapidly turning over (Chapter 11). In the TAT system, both heterokaryons and cell hybrids have been analyzed. In one study, short-term (18–24 hours) heterokaryons of HTC (hepatoma) × rat liver were investigated (Thompson and Gelehrter, 1971). The hepatoma cells synthesize an inducible form of TAT which is not seen in the primary rat liver cells. Both basal level and inducibility were rapidly extinguished in the heterokaryons. Synkaryons have been established on several occasions between HTC × fibroblasts (Table 7.1). The hybrid resembles the fibroblast parent in that it has very low levels of an enzyme that may not be TAT but rather an enzyme with cross specificity for tyrosine.

Weiss and Chaplain (1971) showed that a hybrid that had lost 30–40% of its chromosomes regained enzyme inducibility. Their studies support the Tomkins model in that they show a need to propose two loci: one for the structural gene and one for inducibility. Therefore, inducibility can occur in the absence of appreciable baseline activity.

Croce *et al.* (1973) have proposed that a gene that regulates TAT inducibility is located on the X chromosome. This suggestion is based upon back-selection experiments on somatic cell hybrids using azaguanine, which caused the elimination of one of the X chromosomes and a concomitant reexpression of inducibility of the enzyme. It is interesting to note that the liver-specific enzyme ornithine carbamoyltranferase (EC 2.1.3.3) has been shown to be controlled by an X-linked gene (Fig. 7.2). As previously mentioned, the Lyon hypothesis predicts that X-linked genes will show inactivation in heterozygotes. The X chromosomal inactivation resulting from the Lyon effect was used as a means of determining the location of the ornithine carbamoyltransferase gene in this study. However, this approach does not discriminate between structural and regulatory genes (Ricciuti *et al.*, 1976). Whether or not clustering of genes for liver-specific functions or for regulation in general occurs on the X chromosome remains to be determined.

The observations on TAT activity in hybrids suggest that hypothesis 3 of Davis and Adelberg may explain its regulation in cell hybrids. However, Riddle and Harris (1976) have studied a series of hybrids between HTC cells and

TABLE 7.1 Expression of Differentiated Functions in Liver Cells, Their Derivatives, and Their Hybrids[a]

Marker	Differentiated parental cell	Undifferentiated parental cell	Remarks	References
A. Extinction of differentiated markers in hybrid cells				
TAT, high base-line activity	Hepatoma (rat)	3T3 (mouse)		Schneider and Weiss (1971)
	Hepatoma (rat)	SV40-transformed cells (human)		Schneider and Weiss (1971)
	Hepatoma (rat)	L (mouse)		Schneider and Weiss (1971)
TAT inducibility	Hepatoma (rat)	3T3 (mouse)		Schneider and Weiss (1971)
	Hepatoma (rat)	Epithelial (rat)	Reappearance	Weiss and Chaplain (1971)
	Hepatoma (rat)	3T3 (mouse)		Benedict et al. (1972)
	Hepatoma (rat)	WI-38 (human)	Reappearance linked to loss of human X	Croce et al. (1973b, 1974a,b)
Albumin	Hepatoma (rat)	3T3 (mouse)	Gene dosage effect and cross-activation	Peterson and Weiss (1972)
Aldolase B	Hepatoma (rat)	3T3 (mouse)		Bertolotti and Weiss (1972a)
	Hepatoma (rat)	L (mouse)		Bertolotti and Weiss (1972a)
	Hepatoma (rat)	BRL-1 (rat)	Reappearance	Bertolotti and Weiss (1972b,c, 1974)
	Hepatoma (rat)	DON (hamster)	Reappearance	Weiss et al. (1975)
Aldehyde dehydrogenase	Hepatoma (rat)	L (mouse)	Reappearance	Rintoul et al. (1973a)

Enzyme / Marker	Parental cell	Other cell		Reference
Alcohol dehydrogenase	Hepatoma (rat)	3T3 (mouse)		Bertolotti and Weiss (1972b)
	Hepatoma (rat)	BRL-1 (rat)		Bertolotti and Weiss (1972b)
Alanine amino-transferase	Hepatoma (rat)	BRL-1 (rat)		Sparkes and Weiss (1973)
B. Appearance of new activities not present in parental cells				
Human serum albumin	Mouse hepatoma	Human leukocytes		Darlington et al. (1974a,b)
				Bernhard et al. (1973)
Cross-activation	Rat hepatoma	Mouse 3T3 cells		Peterson and Weiss (1972)
Mouse serum albumin	Rat hepatoma	Mouse lymphoblasts		Malawista and Weiss (1974)
C. Retention of facultative markers in hybrid cells				
Albumin	Rat hepatoma	Mouse 3T3 fibroblast	Cross-activation	Peterson and Weiss (1972)
Transferrin	Mouse hepatoma	Mouse and rat fibroblasts		Szpirer and Szpirer (1976)
Complement (C3)	Mouse hepatoma	Mouse and rat fibroblasts		Szpirer and Szpirer (1976)
Catalase	Rat hepatoma	Mouse L cell		Levisohn and Thompson (1973)
Complement factors	Rat hepatoma	Mouse L cell		Levisohn and Thompson (1973)
Aryl-hydrocarbon hydroxylase	Mouse 3T3 fibroblast	Rat hepatoma		Benedict et al. (1972)

[a] Modified from Ringertz and Savage (1976).

105

different permanent mouse lines, all of which produce some inducible rat TAT. This observation does not agree with a model based on non-species-specific diffusible repressors.

B. Ability to Metabolize Low-Molecular-Weight Intermediates

Because the liver has the ability to synthesize several amino acids from other precursors, hepatic cells of both malignant and nonmalignant origin can proliferate in medium from which these amino acids have been omitted. This provides the basis for a screening system by which the regulation of differentiated gene functions can be studied in culture because undifferentiated cells cannot survive under these conditions. By utilizing reverse selective systems (Chapter 3), it is possible to isolate variants that have now assumed an auxotrophic nature as a result of the loss of one of these liver-specific enzymes. Not only does this represent a unique means by which developmental genetics can be studied, but it may also serve as a model system for the study of metabolic disorders in man, such as phenylketonuria (PKU). This classic inborn error of metabolism results from the absence (PH°) or a drastic reduction (PH⁻) of the liver-specific enzyme phenylalanine hydroxylase (Scriver and Clow, 1980). This enzyme is a component of a complex system for the production of tyrosine from phenylalanine, a system that contains a number of cofactors, the most important of which is $5,6,7,8,$-L-erythrotetrahydrobiopterin (BH_4) and which is the natural cofactor of phenylalanine hydroxylase. Erythrotetrahydrobiopteran transfers protons to phenylalanine and is regenerated by the enzyme dihydropterin reductase. Some cases of hyperphenylalanemia have been found to be due to a deficiency of this latter enzyme. Phenylketonuria is a relatively frequent inborn error of metabolism and results in a catastrophic mental deterioration if left untreated. For this reason it has been the subject of vast newborn screening programs combined with the treatment of affected individuals with a low phenylalanine diet for the first few years of life. Although the exact mechanisms by which the mental retardation associated with PKU is generated is not known, the amelioration of the disorder by dietary therapy is well established.

The observation that the phenylalanine hydroxylating system is present in certain hepatoma cell lines prompted Choo and Cotton (1977, 1979) to isolate through reverse selection a number of mutants unable to survive in tyrosine-deficient medium. Such variants possess between 1 and 20% of wild-type activity (PH° and PH⁻) and are quite stable. By plating cells in tyrosine-deficient medium, revertants can be obtained with a frequency of 10^{-7} to 10^{-6}. The variants in some cases appear to be missense mutations; this conclusion is based on the observation that they may possess an antibody-cross-reacting material. Revertants *always* possess such material. On the basis of this information, it appears most likely that these mutants are the result of alterations in the structural

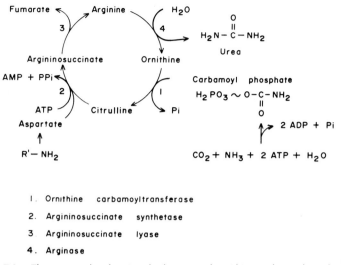

1. Ornithine   carbamoyltransferase

2. Argininosuccinate   synthetase

3. Argininosuccinate   lyase

4. Arginase

Fig. 7.2.   The urea cycle, showing the location of ornithine carbamoyltransferase.

gene for the enzyme, although other possibilities, including mutations in regulated genes, cannot be completely excluded.

Another interesting approach to the study of liver-specific functions has been pursued by Bertolotti *et al.* (1977). Gluconeogenesis is a specialized function confined to the liver. These authors have studied the expression of gluconeogenic enzymes in various liver cell lines using selective medium from which glucose has been omitted. They find that well-differentiated cell lines such as Fu-5-5 (a drug-resistant derivative of HTC) are capable of growth in glucose-free medium whereas other lines are not. This indicates that the differentiated functions required for this process are activated in these cells. Therefore, through the use of forward and reverse selective procedures, it should be possible to isolate rare variants in which genes crucial to differentiation are mutated. This approach appears to be a promising one in the search for regulatory mutants of differentiated gene functions.

A related strategy utilizes enzymes of the urea cycle. Widman *et al.* (1979) have found that rat hepatoma cells fused to normal hepatocytes produce significant levels of the enzyme ornithine carbamoyltransferase and can be selectively grown in arginine-free medium supplemented with ornithine (Fig. 7.2).

C. Additional Liver-Specific Characters

As can be seen from Table 7.1, the situation with regard to the control of gene function in hybrids is contradictory. For instance, Szpirer and Szpirer (1976) found that albumin and α-fetoprotein were extinguished in hybrids, whereas most hybrid clones continued to produce transferrin and the C3 component of complement. This observation suggests that more than one regulatory mechanism exists in liver cells. Aldehyde dehydrogenase, tryptophan pyrrolase, glycogen synthesis, and hepatocyte morphology are all suppressed in hybrids between mouse fibroblasts and fetal liver cells (Rintoul *et al.*, 1973a). In another investigation in which rat hepatoma cells were fused with mouse fibroblasts, the aldehyde dehydrogenase produced after a loss of chromosomes in the hybrid was a new type and possessed different heat inactivation profiles and different isozyme patterns in acrylamide gels (Rintoul and Morrow, 1975). Therefore, it

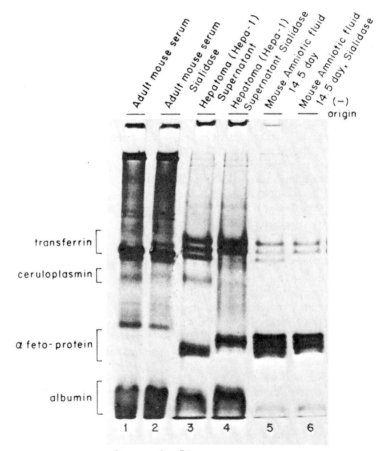

Coomassie Blue

Fig. 7.3. Liver-specific gene functions visualized using polyacrylamide gel electrophoretic separation. (From Darlington *et al.*, 1975.)

appears that the genes for the enzymes in the mouse fibroblast parent were reactivated by the presence of the rat hepatoma genes.

Another cell line of liver origin is the mouse hepatoma (Hepa 1a) which possesses a number of differentiated liver functions (Fig. 7.3), including serum albumin and α-fetoprotein production. When Hepa 1a cells were fused with human leukocytes, a number of human genes products were demonstrated in the hybrids. Mouse *and* human albumin are produced, implying that the human structural gene for albumin was activated. Szpirer *et al.* (1980) have studied the production of several liver proteins in mouse hepatoma–adult rat hepatocyte hybrids. Such hybrids appear to have lost the majority of their rat chromosomes, an observation that is in agreement with previously cited data on fusions between rapidly and slowly dividing cell lines. All the hybrids produce mouse albumin, transferrin, and C3 protein, and most clones secreted at least one rat serum protein. In none of the clones was the rat gene for α-fetoprotein activated (it is not produced in normal adult liver); however, neither did the rat genome suppress the mouse α-fetoprotein production. These observations indicate that fusion of a normal, nondividing cell with a tumor cell of the same tissue type can allow the "immortalization" of the functions of the former cell type. This appears to be a common observation in the study of differentiated hybrids (see Chapters 8 and 9).

III. CLONAL VARIATION IN LIVER CELLS

Wide variations in aryl-hydrocarbon hydroxylase have been noted in different subclones in both diploid and heteroploid lines (Whitlock *et al.,* 1976). In this case, large alterations in karyotype appear *not* to be involved. Hankinson (1979) has measured the rate of loss of this enzyme by isolating clones resistant to benzo[*a*]pyrene. Because aryl-hydrocarbon hydroxylase converts benzo[*a*]pyrene to a toxic intermediate, resistance can be achieved through loss of this enzyme. Using the fluctuation test, a variation rate of 2×10^{-7} per cell per generation was recorded. In this case, the low rate suggests loss of the enzyme due to mutation. Although the previously cited study makes a convincing case for mutation in structural genes as the basis of variation for liver-specific characters, there may be other mechanisms that produce such quantitative alterations. If hepatoma cells are plated out and many individual clones evaluated for quantitative enzyme levels, large variations are encountered (Table 7.2) (Bernhard *et al.,* 1973). This wide clonal variation in amount of gene product synthesized has been studied in detail by Peterson (1974, 1979). In the case of albumin production, clonal variation in hepatoma cells and hepatoma–fibroblast hybrids encompasses a 200-fold range. The rates of variation, established by the Luria–Delbrück fluctuation test, were extremely high: $0.5–1.4 \times 10^{-2}$ per cell per

TABLE 7.2   Serum Albumin Secretion Rates of Cultured Mouse
Hepatoma Cells[a,b]

Cell line	Molecules per cell per minute
Hepa	57,600
Hepa 1[c]	33,600
1a[c]	18,400
1b	176,000
1c	80,000
1d	72,000
1e	56,000
1f	44,000
1g	36,000
1h	20,800
1i	20,000
1k	18,400

[a] Mouse serum albumin was accumulated in the supernatant medium and assayed by electroimmunodiffusion. Molecules secreted per minute per cell are calculated by assuming a molecular weight of 69,000 for mouse serum albumin.

[b] Modified from Bernhard et al. (1973).

[c] Means and standard deviations are calculated from three experiments, which were performed over a period of 4 months.

generation. This is several orders of magnitude higher than most cell culture "mutation" rates (see Chapter 1) and suggests a completely different mechanism. The variation is not random but follows a clustered geometric progression. Furthermore, the amount of albumin messenger is proportional to the quantity of protein and messenger synthesized. The significance of these observations to the control of differentiation is obscure, but they suggest (1) redundancy of either the structural gene or the regulatory gene responsible for the character, or (2) alteration in rate of transcription through a promoter mutation.

IV. TRANSFER OF CONTROL FACTORS VIA CYTOPLASMS

In two rather surprising reports, liver-specific functions suppressed by fusion with fibroblasts were activated when cytoplasms of hepatoma cells were fused to fibroblastic cells (Lipsich et al., 1979; Gopalakrishnan and Anderson, 1979). The first report concerns TAT and the second concerns phenylalanine hydroxylase. One could explain these results by postulating an epigenetic cytoplasmic inducer of specific liver function which is produced in such limited quantities that all of it is utilized in the induction of hepatoma cells. In the presence of both the L cell and the hepatoma genomes (cell hybrids) the hypothetical inducer is diluted to a point where it is ineffective in bringing about the synthesis of the liver-specific function in either genome.

A cytoplasmically controlled factor would also explain the observation that when tetraploid hepatoma cells (2s) were fused with diploid (1s) fibroblasts, the hybrids synthesize four liver-specific enzymes that are suppressed in 1s × 1s hybrids. These are TAT, alanine aminotransferase, aldolase B, and alcohol dehydrogenase (Brown and Weiss, 1975).

Such a model of differentiated regulation would require that the cytoplasmic factors responsible for regulation be present as a fixed number of doses, that they replicate at the same rate as the cell in which they reside, and that they be duplicated in polyploidized cells. Furthermore, by virtue of their cytoplasmic nature, they would not be localizable to particular chromosomes through cell hybrid mapping strategies. Because the experiments using polyploidized, differentiated cells argue that these hypothetical factors are titratable, they would have to be present in a number of copies rougly equal to the number of genes that they suppress, which would suggest a handful of molecules per cell.

V. CONCLUSION

At present we do not have a comprehensive picture that would explain the mechanism by which the intermittent synthesis of gene products is controlled in liver cells and their derivatives. In somatic cell hybrids almost every possibility of expression has been encountered: suppression, continued functioning, or activation of previously quiescent genes. Nor is the basis of variation of liver-specific functions understood. Although a high rate of variation for albumin production has been encountered in rat and mouse hepatoma cells, suggesting a nonmutational basis for this phenomenon, some other differentiated systems (aryl-hydrocarbon hydroxylase) have such low mutation rates to nonproduction that one suspects a mutation in either a structural or a regulatory gene to be responsible.

One possible pattern that apparently has not been encountered is that of simultaneous expression to different histiotypes in the same cell. Fougere and Weiss (1978) have noted that in melanoma–hepatoma hybrids melanin production and albumin production could not be expressed in the same cell, even though the phenotypes appeared to switch on and off in the process of subcloning. The authors note that this observation is in agreement with the proposition of ''mutual exclusion'' put forth many years ago by Weiss (1939). The fact that this generalization of the discreteness of differentiation applies to cell hybrids would indicate that they follow the rules of normal development and are therefore useful material for the study of differentiation.

One generalization that can be made at this time concerns the molecular control of albumin production. In a case studied by Leinwand *et al.* (1978), the levels of specific messenger RNA for albumin were proportional to the amount of albumin synthesized in parents and hybrids between rat hepatoma × L cell hybrids. This suggests that at least in the case of albumin production, control is regulated at the transcriptional level. Whether or not this rule applies to other characters and what its precise mechanism may be remains for the future to decide. It is certain that gene transfer experiments for differentiated gene functions will be useful in resolving this question.

8

Differentiation in Cultured Cells: Muscle Cells, Melanoma Cells, Neuronal Cells, and Hemoglobin-Producing Cells

I. INTRODUCTION

In addition to normal and malignant cells from the liver, a number of differentiated cell types have been investigated in culture systems. Many of these, such as cells derived from bone, fat, and pancreas, have not been considered experimentally from the standpoint of somatic cell genetics and will not be reviewed here. However, four cell types that have been utilized in hybridization systems have illuminated mechanisms of gene regulation *in vitro*. Furthermore, they are particularly of interest because it appears that they may be used as model systems for several important human disorders that have a known or possible genetic basis. They will be dealt with as a unit because the approach to their study and many of the findings concerning their genetic properties are similar.

II. MUSCLE

There has been no dearth of interest in this tissue as a result, no doubt, of several factors:

1. A wide assortment of extremely debilitating diseases are expressed in muscle. The most widespread are muscular dystrophies, of which Duchenne muscular dystrophy is the most frequent. This is a chronic, wasting, degenerative

disorder that is the result of a sex-linked recessive mutation. Whether it is due to a genetically determined defect in muscle or in the circulation, or to an abnormal influence of neuronal origin upon the muscle is not known at present. Many biochemical studies have compared normal and Duchenne muscle tissue; however, most of the results have shown that the properties of the tissues are the same or that differences in the Duchenne muscle can be ascribed to secondary effects. One exception to this was the observation that proteins of the sarcoplasmic reticulum are modified (Samaha and Nagy, 1982). These disorders could perhaps be understood on a cellular level through the use of primary and permanent myoblasts, the precursors of highly differentiated muscle cells.

2. Muscle cells produce a variety of well-characterized gene products including actin, myosin, creatine phosphokinase, and other proteins. Furthermore, both permanent myoblast cell lines and primary muscle cell cultures undergo a fusion process that results in a multinucleate syncytium as the cultures reach confluency (Fig. 8.1).

3. Muscle is a tissue that lends itself easily to experimental manipulations, including establishment of primary cultures and permanent lines and isolation of cell hybrids and cybrids. In a cell culture environment it can be removed from

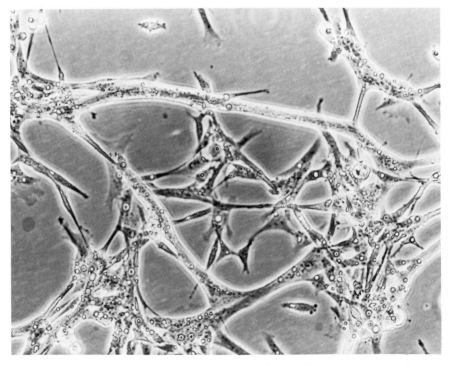

Fig. 8.1.   Quail myoblasts undergoing fusion in culture with multinucleate myotubes.

contact with elements of the nervous system and questions relating to the neurogenic or myogenic origin of particular disorders can be explored.

Perhaps the central question in this system is, By what mechanism does differentiation of the myoblast occur (Harris, 1980)? This pattern of activity occurs when either primary cultures of avian or murine origin (Coleman and Coleman, 1968) or established rat myoblasts (Shainberg *et al.*, 1971) reach confluence and undergo fusion (Fig. 8.1). This transition is accompanied by a complex series of molecular events. Prior to fusion there is a decrease in DNA polymerase activity, DNA synthesis, and the rate of cell division (Zallin and Montague, 1974). Subsequent to this there are striking changes in protein synthesis, including a several-hundred-fold increase in the synthesis of actin, myosin, and related components of the sarcomeres which can be identified with the aid of the electron microscope (Rubenstein and Spudich, 1977). There is a large increase in the level of creatine phosphokinase as a result of the synthesis of the M or muscle isozymic form, which replaces the B or brain form (Dym *et al.*, 1978) that is predominant prior to fusion. These isozymic changes are of particular interest in the study of muscular dystrophy, because in this disorder the B form of the enzyme is found in adult muscle, both in human (Kuby *et al.*, 1977) and avian (Weinstock *et al.*, 1978) dystrophic cells. In normal adult muscle, no B form is found. Furthermore, striking changes in cell morphology, in particular fusion of the myoblasts to form pronounced myotubes, can be seen; these changes result in the formation of terminally differentiated muscle fibers. Additional changes at the cell surface include increases in acetylcholinesterase activity and in acetylcholine receptors (Fluck and Strohman, 1973) and a transition of fibronectin from a fibrillar pattern to one of discrete clusters (Chen, 1977). The investigation of these events at the cellular and molecular level required the development of methods for the isolation and culturing of muscle cells. Such procedures have now been perfected and are similar to those involved in the handling of many other tissues (Thompson, 1980). Fragments of muscle are gently dissociated mechanically and, through the use of proteolytic enzymes such as collagenase, are broken into cell suspensions. Tissue from fetal animals ordinarily is more successful for the development of myoblasts *in vitro,* as is true for most differentiated material. Cells are plated into a specially formulated medium and are treated to remove fibroblastic elements, which otherwise will overgrow the myoblasts. This may be accomplished through the use of maintenance medium and through serial transfer of the cells (Askanas and Engel, 1975), a process by which the muscle tissue is purified of fibroblasts by several transfers over a period of weeks.

Over the years a number of permanent myoblast cell lines that have been developed have retained many of their differentiated features (Sanwal, 1979). Because primary muscle cultures do not proliferate continuously, the permanent myoblast lines have been quite useful in resolving questions of a biochemical nature, including the study of the fusion process itself (Rogers *et al.*, 1978). One of the most widely employed myoblast lines is the L6 line originated by Yaffe (1968). This culture was obtained from the skeletal muscle of the rat and retains the ability to form myotubes, a process that is accompanied by the synthesis and accumulation of muscle protein (Shainberg *et al.*, 1971).

The relationship between the morphological and biochemical aspects of fusion has been investigated with the aid of fusion-negative mutants (Sanwal, 1979). Such mutants can be easily obtained in L6 and L8 cell lines; one thoroughly studied case was observed to produce creatine phosphokinase but not myosin, heavy chains, or acetylcholine receptors (Rogers *et al.*, 1978). The expression of some, but not all, of the differentiated features of the myoblast demonstrated that they are not regulated as a group. This conclusion is bolstered by an analysis of cell hybrids between fibroblasts and myoblasts; in these hybrids, fusion can occur in the absence of differentiation.

Another fusion-negative variant of the L8 cell line was studied in detail by Kaufman and Parks (1977). Such cells behave as transformed cells: that is, they grow in soft agar and cause tumors in nude mice. Furthermore, they have lost many of their differentiated properties, including the increased synthesis of creatine phosphokinase and myosin that is normally observed upon fusion. These observations suggest that the L8 myoblast cell line is a terminally differentiated cell that retains most mechanisms of normal growth and development.

In another report (Dufresne *et al.*, 1976), a number of muscle-specific properties were suppressed in myoblast–fibroblast hybrids, including fusion and synthesis of creatine phosphokinase and myosin. These data are in agreement with the previously cited studies and with the usual observation that most differentiated characters are suppressed in cell hybrids (see Chapter 7).

Ringertz *et al.* (1978) have generated reconstituted cells from rat myoblast karyoplasts and mouse fibroblast cytoplasts; both were produced through the use of cytochalasin B. Such cells retain their myogenic phenotype, indicating that the program of muscle-specific gene expression is not interrupted by contact with the fibroblast cytoplasm.

An alternative approach to the study of muscle cell differentiation is the use of chemical agents that specifically affect its expression. The use of BUdR will be discussed in a subsequent chapter; another agent that has been considered is dimethyl sulfoxide (Blau and Epstein, 1979). At a final concentration of 2% in culture medium, this substance prevents the fusion of myoblasts and the accompanying increase in creatine phosphokinase and acetylcholine esterase activities. In addition, synthesis of actin and acetylcholine receptor protein is not seen. The mode of action of DMSO is not understood at present. Because it is a bipolar substance, it may have effects on membranes that indirectly affect the differentiation process. In contrast to the coordinate control seen under the influence of

DMSO, cytochalasin B prevents fusion while allowing the parameters of differentiation to proceed unimpeded. These studies are in agreement with other investigations that indicate that fusion and biochemical differentiation are not inextricably linked (Holtzer *et al.*, 1972). Thus, fusion is neither a prerequisite nor an initiator of myogenesis.

The development of recombinant DNA technology has made possible the study of differentiation in the muscle cell at the molecular level (Jones, 1980). Furthermore, these techniques offer the hope that the genetic defects responsible for muscular dystrophy could be uncovered by working backward from the gene to the product. An obvious starting point for such investigations is the cloning of the myosin gene. Because myosin is an extremely large protein (200,000 MW), it is generated from a large translational complex that is relatively easy to purify. From the polyribosomal complex, a pure myosin heavy chain mRNA has been isolated and has been employed to direct the synthesis of a complementary DNA (cDNA). This cDNA behaves as a muscle-specific molecule, hybridizing only with mRNA from myogenic cells. Jones goes on to discuss the possibility of human gene cloning through the use of human–mouse hybrids that retain only the human X chromosome. By generating a library from the DNA of such a cell line, it would be possible to test for hybridization with mRNA from myogenic human cells. Although such a procedure might uncover the existence of 200 or more muscle-specific, X-linked genes, it would open the door to the possibility of isolating the gene responsible for muscular dystrophy. Such an X chromosome-specific library has already been constructed by Davies *et al.* (1981), who used an innovative approach. These researchers obtained a 48-chromosome, tetra-X fibroblast cell line (48, XXXX) from the Genetic Mutant Cell Repository and separated the X chromosomes on a fluorescence-activated cell sorter. Such a device uses laser identification to identify and separate biological entities, which can be utilized for the construction of an X chromosome-specific library.

An effort has recently been made to account for myoblast differentiation on the level of transcription and translation and in the patterns of particular proteins synthesized (Affara *et al.*, 1980). In these investigations, cells from three different stages of myogenesis were compared: myotubes formed *in vitro*, myoblasts derived from a mouse teratocarcinoma, and an undifferentiated mouse embryonic carcinoma. It was found that new DNA sequences were transcribed in the myotubes and myoblasts, as evidenced by a 30% increase in hybridization of total mRNA to single-copy mouse DNA. Studies on the cytoplasmic mRNA using cDNA probes from each of the three stages of differentiation demonstrated that at each level a new family of messengers enters the polysomes. Furthermore, the use of cDNA probes enriched for myotube sequences indicates that the mRNA appearing in the differentiated cell is absent or much reduced compared to the mRNA of undifferentiated cells. This establishes that regulation is at the level of RNA synthesis or degradation, a conclusion substantiated by a similar study by Shani *et al.* (1981). Furthermore, translation *in vitro* of purified messages showed that the polypeptides synthesized from the abundant mRNA class were the same as those synthesized by myotubes *in vivo*. These polypeptides include myosin light chains, muscle actin, and muscle troponin.

If differentiation in myoblast cells is controlled at the level of transcription, by what mechanism might synthesis of appropriate messengers be initiated? An interesting possibility has arisen from the work of Jones and Taylor (1980), who have demonstrated that the nucleoside analog 5-azacytidine causes hypomethylation of DNA and also induces myotube formation in mouse embryo cells. These authors propose that the formation of 5-methylcytosine in DNA in the sequence CCGG results in an alteration of the DNA that is preserved in subsequent rounds of replication. The possible role of hypomethylation in gene regulation has been stressed for a number of systems, including activation of the mammalian X chromosome (Chapter 13). Although the nature of the evidence is at present indirect, it is highly suggestive of a role of hypomethylation in gene regulation.

III. MELANIN SYNTHESIS IN CELL HYBRIDS

Vertebrates possess a specialized cell type for the production of pigment; this cell is known as the melanocyte. Within this cell a complex of molecules is produced and forms the brown pigment melanin. This occurs by way of a series of reactions in which tyrosine and/or dihydroxyphenylalanine (DOPA) is converted to DOPA quinone by DOPA oxidase. The subsequent oxidation and polymerization results in the conversion of the DOPA quinone to a material that is associated with a protein framework and becomes melanin. Immature pigment granules are known as premelanosomes, and they eventually mature into melanosomes (Brumbaugh and Lee, 1975).

Hybrids have been produced between the Syrian hamster melanoma cell line and a number of nonpigmented cell types. In a large number of such hybrids it was observed that pigment synthesis was extinguished (Fig. 8.2). Furthermore, the enzyme DOPA oxidase is not produced in such hybrids, as is the case for the parental nonpigmented cell lines. The pigmented Syrian hamster parent, however, produced easily detectable quantities of this enzyme (Davidson, 1974). Such experiments suggested the presence of a diffusible regulator that suppresses synthesis of the enzyme and, thus, pigment production. Subsequently, a polyploid melanoma cell line was developed that was assumed to carry two pigment cell genomes. When such cells are fused with unpigmented fibroblasts, 50% of the hybrids were pigmented, in striking contrast to the results obtained with the 1s × 1s hybrids. As in the previous series of hybrids, the DOPA oxidase activity paralleled the appearance of pigment. Pigmented clones segregated out unpig-

Fig. 8.2. Melanin-producing phenotypes in parental lines and production somatic cell hybrids. (Left to right) Melanoma parent, mouse fibroblasts, 1s melanoma × 1s fibroblast, 2s (tetraploid) melanoma, unpigmented 2s × 1s hybrid, pigmented 2s × 1s hybrid. (From Davidson, 1974.)

mented daughter clones, but unpigmented clones were never observed to yield pigmented progeny.

More recently, Halaban *et al.* (1980) have published results of hybridization experiments between normal mouse embryo fibroblasts and Cloudman melanoma cells. These hybrids are pigmented and produce substantial quantities of tyrosinase, the enzyme whose absence is responsible for albinism. At first inspection, these results would appear to be contrary to the findings of Davidson (1974); however, the Cloudman melanoma is a pseudotetraploid, and it is quite likely that the hybrids possess two or more doses of the melanin-producing genes. Thus, these hybrids may be analogous to the 2s/1s hybrids that are described by Davidson and that produce melanin and DOPA oxidase.

The observance of melanoma–fibroblast hybrids containing different doses of the pigment-producing genome suggests that repressor molecules are produced by the undifferentiated parent in limited quantities. According to this hypothesis, the double dose of differentiated genes in the 2s/1s hybrids dilutes out the repressors produced by the mouse fibroblast below their level of optimum effectiveness (see Chapter 7).

IV. NERVE CELLS

The nervous system is a complex system of circuits; it stores and transmits information through the use of integrated collections of cells that are connected by fluid-filled processes known as axons. Signals are transmitted electrically by polarity changes down the length of the axon, while at the same time these processes serve to transport nutrients to and from the cell body. The study of cultured cells from the nervous system offers the opportunity to answer many questions regarding mechanisms of differentiation, the establishment of connections between cells, and the nature of messages that are sent and received between nerve cells.

The major cell type of the nervous system is the neuron. In addition, accessory cells including glia and astrocytes are found throughout, and other contributing elements, such as Schwann cells, are intimately related to the function of the brain. An understanding of the means by which these elements are integrated into a functioning whole would contribute to our understanding of the means by which information is stored, processed, and transferred by the brain.

In this tissue, as in others, permanent cell lines that are capable of producing differentiated products while still retaining the ability to divide are of great value. Not only do they (at least in some cases) retain many of the proteins characteristic of their tissue of origin, but the fact that they can be grown in pure culture means that functions can be unambiguously assigned to a particular cell type. By using cells derived from tumors of the nervous system, it has been possible to develop clonal lines from both glia and neurons (Rosenberg, 1973); such cell lines have been shown to possess many of the biochemical properties of the

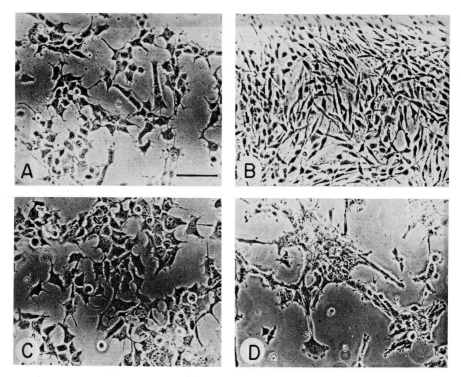

Fig. 8.3. Representative neuroblastoma (A), glioma (B), and neuroblastoma–glioma hybrids (C and D). (From Amano et al., 1974.)

TABLE 8.1   **Differentiated Characters Expressed in Cells Derived from the Central Nervous System and Their Hybrids**

Cell type	Property	Reference
C1300 neuroblastoma	Acetylcholinesterase	Yogeeswaran et al. (1973)
	High steroid sulfatase in confluent cultures	McMorris et al. (1974)
	Choline acetyltransferase	Prasad and Mandal (1973)
	6-Hydroxydopamine sensitivity	Prasad (1971)
	Enzymes for nor-epinephrine synthesis	Schubert et al. (1969)
	14-3-2 protein	Yogeeswaran et al. (1973)
	Electrical excitability	Nelson et al. (1969)
	Neurite extension	Schubert et al. (1969)
	Tyrosine hydroxylase	Waymire et al. (1972)
Rat glioma	S100 protein	Benda et al. (1968)
	Glial-specific hormonal induction of glycerol-3-phosphate dehydrogenase	DeVellis and Inglish (1969)
Neuroblastoma × glioma hybrid	Morphine receptors	Klee and Nirenberg (1974)
	Synthesize acetylcholine	Daniels and Hamprecht (1974)
	Synapse formation with muscle cells	Nelson et al. (1976)
	Choline acetyltransferase	Amano et al. (1974)
Neuroblastoma × neuron hybrid	Electrical excitability	Chalazonitis et al. (1975)
Neuroblastoma × mouse fibroblast	Electrical excitability	Minna et al. (1971, 1972)
	14-3-2 protein	McMorris et al. (1974)
	Acetylcholinesterase	Minna et al. (1971, 1972)

normal nervous system (Rosenberg *et al.*, 1978). Tumor cells derived from neurons are referred to as neuroblastomas, and they possess such qualities as the ability to extend long axon-like processes (Fig. 8.3A), whereas glial cell tumors (gliomas) can synthesize a brain-specific protein known as S100 (Fig. 8.3B). The many properties assigned to these cells and to representative hybrids (Fig. 8.3C and D) are listed in Table 8.1.

It will be noted from Table 8.1 that, in contrast to many of the other differentiated cell lines, neuroblastoma × fibroblast hybrids often express as much of or more of the differentiated product than does the NB parent cell line. Furthermore, it is possible to produce neuron–neuroblastoma hybrids in which the properties of electrical excitability more closely resemble those of the neuronal parent than those of the neuroblastoma parent (Chalazonitis *et al.*, 1975).

A complex pattern of intercellular interaction involving elements of the nervous system has been demonstrated using an *in vitro* model system. Neuroblastoma–glioma hybrids form chemical synapses with striated muscle cells (Nelson *et al.*, 1976), as judged by the elicitation of muscle responses following action potential formation in the hybrid neuronal cells. Although synapse formation occurs with ease, there is no molecular mechanism known at present that would facilitate these connections. Presumably, specific recognition mechanisms that would enable specific nerve cells to join to certain muscle fibers are operative.

One of the most exciting areas of investigation in neurobiology concerns the mode of action of opiates. Opiates are endogenous or synthetic peptides that have a spectrum of pharmacological activities similar to morphine (Beaumont and Hughes, 1979). Opiates (or enkephalins) have been widely implicated as neurotransmitters or neuromodulators at the synaptic junction. With the goal of developing an *in vitro* system for the consideration of this problem, investigators tested neuroblastoma–glioma cell hybrids for specific binding of ³H-labeled morphine (Klee and Nirenberg, 1974). Although the parent cell lines had few or no receptors, it was observed that one of the hybrid clones displayed high-affinity morphine receptors. The hybrids also possessed several neuronal properties, such as high levels of choline acetyltransferase (Amano *et al.*, 1974); this finding indicated that the hybrids could be quite useful for a variety of studies on

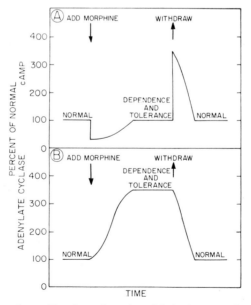

Fig. 8.4. Pattern of morphine dependence in cell hybrids. (From Sharma *et al.*, 1975a.)

the action of narcotic analgesics. Such receptors have also been demonstrated in neuroblastoma cells by using a fluorescent analog of enkephalin; the receptors are distributed in clusters (Hazum *et al.*, 1979).

A related study by Sharma *et al.* (1975a,b) has concerned itself with the mechanism of narcotic dependence because after 1 to 2 days the inhibition and increase balance one another and the level of the enzyme returns to normal. Upon withdrawal of morphine, however, the inhibition is removed and levels of the enzyme are abnormally elevated, which is similar to morphine dependence in animals (Fig. 8.4).

The expression of neuron-specific properties is associated with changes in the species of messengers produced (Grouse *et al.*, 1980). When neuroblastoma cells are stimulated to differentiate with dibutyryl cyclic AMP, a class of messenger not present in undifferentiated cells appears. The molecular basis of this phenomenon may be similar to that of the myoblast differentiation described earlier.

V. HEMOGLOBIN SYNTHESIS IN CELL HYBRIDS

The study of hemoglobin has occupied a pivotal role in the development of genetics. Its analysis has resulted in such a host of fundamental discoveries that a substantial textbook of genetics could be written on the basis of this system alone. From the time in the late 1940s when sickle cell anemia was recognized as a "molecular disease" (Pauling *et al.*, 1949) up to the present, every aspect of the genetics of man has been touched in a profound manner by these investigations. An understanding of its significance to eukaryotic genetics requires some consideration of the diverse approaches of this research.

In humans, hemoglobin, the oxygen- and carbon dioxide-transporting pigment in blood, consists of four globin chains; each chain is attached to a heme group which binds the oxygen and the carbon dioxide molecules. Two of these chains are usually α chains, designated α_2, and each α chain consists of over 140 amino acids. The pair of α chains can combine with a variety of other chains which are indicated by different letters of the Greek alphabet. Thus, the α chains associate at different stages of development with other different chains to form the quaternary structure of the hemoglobin molecules. Each of the different globin polypeptides is produced by a different gene, but their amino acid sequences are so similar as to indicate a common evolutionary origin through gene duplication. Hemoglobin A is the most common type, and it consists of two α and two β chains. This is the predominant adult type, although a minority (2 to 3%) type known as hemoglobin A_2 is also present ($\alpha_2\delta_2$). Several types of hemoglobin are produced in very early life: $\alpha_2\gamma_2$ or hemoglobin F (fetal), $\alpha_2\epsilon_2$ or hemoglobin E (embryonic), and zeta (ζ) chains, which are the earliest produced (Fig. 8.5).

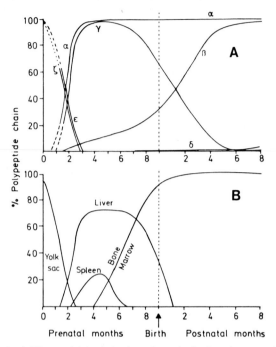

Fig. 8.5. Levels of different globin chains in embryonic, fetal, and postnatal life. (A) Proportions of various molecular species; (B) relative participation of various blood-forming organs. (From Vogel and Motulsky, 1980.)

Restriction endonuclease mapping has shown that two of the globin genes appear to have been duplicated in the recent evolutionary past (Orkin, 1978). In most humans the α gene appears in two tandem, identical copies on chromosome 16, whereas the γ gene occurs in two copies that differ only on the basis of a single amino acid residue (position 131; γ_a = alanine; γ_g = glycine) and that are also located in tandem. A wealth of genetic variation occurs in the hemoglobin genes, including point mutations, deletions, and duplications. Furthermore, mutants that have arisen by unequal crossing over have resulted in fusions of some of the genes; this process has enabled the establishment of linkage relationships between the δ and β and the γ and β genes. These genes form a complex located on chromosome 11 (Deisseroth et al., 1978).

A recent analysis of the α globin gene in the mouse (Leder et al., 1981) has established the existence of a complex, multigene family. This includes an embryo α gene and two adult α genes, all three of which are located within a 25-kb fragment on chromosome 11. However, the recombinant DNA analysis employed here established that two additional α "pseudogenes" were present on different chromosomes. These nonfunctional pseudogenes appear to arise from active genes by an amplification event and may undergo subsequent genetic drift resulting in their inactivation and sequence diversification.

Many of the mutations within the human hemoglobin molecule are associated with resistance to malaria. Individuals heterozygous for such genes are at a selective advantage in a malarial environment despite the extreme debilitation brought about by some of these anemias in individuals in the homozygous state. The study of these conditions has contributed a great deal to our understanding of human evolution, especially investigations on sickle cell anemia. This extremely serious anemia is the result of a mutation in the β gene (valine for glutamine at position 6). It is perhaps the only genetic variant for which a heterozygous advantage has clearly been demonstrated (Vogel and Motulsky, 1980).

Sickle cell anemia is widespread among black populations throughout the malaria-infested regions of the world and reaches frequencies ranging from 25 to 40% for the heterozygote. A wealth of evidence has established the relationship between the high gene frequencies and resistance to falciparum malaria among the heterozygotes. This evidence includes the extremely high childhood mortality in endemic areas, the loss of the gene from black African populations transplanted to nonmalarial regions, and experiments *in vitro* showing that the red cell from sicklers is a less adequate growth environment for the malarial parasite. Thus, even though the homozygous sickle cell condition can be 100% fatal, it is under extremely strong selective pressure, especially in regions whose ecology has been modified to favor the breeding of the mosquito.

Another class of disorders affecting hemoglobin comprises the thalassemia diseases, in which the quantity of hemoglobin is diminished. Although it was originally suspected that these disorders might be the result of mutations in operator or regulator genes, it is now clear that deletion of α genes in increasing numbers is responsible for the more serious cases of α-thalassemia. Restriction mapping of DNA from individuals with different degrees of α-thalassemia has demonstrated that whereas in the normal individual the α genes are present as two identical tandem duplications, deletions of up to three of the four copies normally present in a diploid genome result in the defective synthesis of the polypeptide chains that characterizes this disorder. β-Thalassemia, a somewhat less serious disorder, is not yet completely analyzed at the molecular level. However, in cases of the disease β^0-thalassemia, the β gene is present but no β chains are produced. This situation may represent a regulatory mutation (Nienhuis and Benz, 1977).

It is evident that a system so well understood at so many levels should lend itself in a most exciting manner to cell culture investigations on the mechanism of gene regulation. Although theoretically this system could be studied using primary cultures of erythroid cells, malignant cell lines capable of making hemoglobin have, in fact, proved to be of greater value. Virus-transformed erythroid precursor cells can be induced to synthesize hemoglobin when treated with dimethyl sulfoxide (Nudel *et al.*, 1977), butyric acid (Leder and Leder, 1975), or organic solvents, and such cells have allowed investigators to study the activation and expression of the globin genes. Furthermore, the production of cell hybrids between mouse erythroleukemia cells and human erythroid cells has provided a method by which human genetic disorders of hemoglobin production could be investigated in culture.

Murine erythroleukemia cells are characterized by a number of features shared by differentiated erythroid cells. These include (1) erythrocyte membrane antigen, (2) increase in heme synthesis and iron uptake (Friend *et al.*, 1971), (3) induction of hemoglobin synthesis (Ikawa *et al.*, 1979), and (4) the accumulation of globin mRNA (Ross *et al.*, 1972). The most straightforward fashion in which to detect differentiation in these cells is through a simple stain for hemoglobin. The benzidine staining method (Orkin *et al.*, 1975) is simple and can be quantified through the use of microspectrophotometry.

Regulation of genes controlling hemoglobin production can be brought about through the use of chemical agents or by cell hybridization. As in other differentiated systems, the production of globin is suppressed in cell hybrids derived from fusions of fibroblasts and hemoglobin-producing cells. This suppression extends to the manufacture of globin mRNA, including suppression of both constitutive production and induction in the presence of DMSO. The results of earlier experiments suggested that the induction of erythroid differentiation by DMSO requires exposure during the S phase of the cell cycle (Levy *et al.*, 1975). However, the results of more recent experiments using cells synchronized by flow microspectrofluorimetry indicate that the commitment to differentiation and globin mRNA synthesis are coupled and that both events occur in G_1 phase following a precommitment phase of 12 hours. By using specific molecular probes, it has been shown that the structural genes for globin production are present in these hybrids (Deisseroth *et al.*, 1976).

Analysis of hemoglobin production by nonproducing cell hybrids indicated that a gene on the X chromosome of the undifferentiated parent suppresses hemoglobin synthesis (Benoff and Skoultchi, 1977). Further analysis (Benoff *et al.*, 1978) indicated that the mode of action is through the suppression of heme synthesis by inhibition of the step catalyzed by levulinic acid synthetase.

In a subsequent analysis of mouse erythroleukemia × human fibroblast hybrids (Wiling *et al.*, 1979) it was observed that human globin genes could be activated provided that two doses of the erythroleukemia genome were fused with one human fibroblast genome. In a related system it was shown that the tetraploid erythroleukemia cells fused to teratocarcinoma cells expressed hemoglobin genes from both parents (McBurney *et al.*, 1978). These findings imply the existence of positive regulatory factors; however, the activation is selective. Both human α and human β globin genes were activated; however, the γ gene (fetal hemoglobin) was not, even though it is closely linked to the β gene.

Because the α and β genes are on separate chromosomes, the regulatory system must be trans active (i.e., able to influence genes on other chromosomes).

Finally, human globin genes could be activated by chromosomal transfer of the globin gene from human mononuclear peripheral blood cells into tetraploid erythroleukemia cells. This observation further supports the contention that diffusible, trans-acting factors will activate genes (under specified conditions).

VI. CONCLUSION AND SUMMARY

In this chapter we have considered several tissues that express a variety of differentiated properties in culture and that may lend themselves to the understanding of fundamental questions regarding disease states, in addition to expanding our analysis of differentiation on a molecular and cellular level.

The most important disorders that affect the muscle cell are the dystrophic diseases, and although we as yet have no notion what the cellular basis of these conditions results from, important developments in the field of muscle cell culture are encouraging, because they suggest that *in vitro* systems will shortly be available to aid in our understanding of these diseases. The cloning of muscle-specific genes through recombinant DNA technology should be especially valuable to our understanding of normal and pathological function of muscle cells.

In the field of neurobiology, glioma and neuroblastoma tumors have been quite important in the development of *in vitro* models of the nervous system. The surprising observation that cell hybrids appear in some cases to produce greater quantities of tissue-specific gene products than do the parent lines has proved to be important in the formulation of models of narcotic addiction and dependency. Thus, the presence of specific morphine receptors on the cell surface of the hybrid is a valuable tool in the study of the neuropharmacology of these substances.

Recombinant DNA technology is presently being employed to further our understanding of globin gene regulation in hemoglobin-producing cell lines and their cell hybrids. By using specific probes for the globin genes it is possible to demonstrate the presence of the genes in cell hybrids even though the genes are in a nonfunctional state. This observation would argue for the existence of positive regulatory factors produced by the undifferentiated parent.

9

Differentiation in Cultured Cells: Cells of the Immune System

I. INTRODUCTION

Stunning progress has characterized the study of immunogenetics in recent years. In the 1960s, hypotheses were outlined to account for the manner in which cells of the immune system were capable of producing specific antibodies that complex with any of the myriad molecular forms capable of behaving as an antigen through a lock-and-key type mechanism. During the 1970s these models were tested by using myeloma tumor cells that can be grown in culture and that produce monomolecular species of antibodies. Myeloma tumor cells are malignant, transformed plasma cells that have been widely investigated in a number of species, including mice, rats, and humans. They produce immunoglobulin molecules of various levels of complexity. Because they can be cultivated either *in vivo* or *in vitro* and because they produce a single, homogeneous molecular species, they have been crucial in the development of our understanding of immunology. Through the use of the myeloma cell lines, large quantities of homogeneous antibody molecules can be isolated and purified, allowing a complete biochemical description of these entities. Presently molecular probes of cloned immunoglobulin genes are being employed in order to elucidate the manner in which the genes are arranged along the chromosome and the consequences of this architecture for the expression and control of immunoglobulin synthesis.

II. MECHANISM OF ANTIBODY DIVERSITY

When a bone marrow-derived lymphocyte is stimulated by the presence of an antigen, it differentiates into a plasma cell and begins to produce a specific

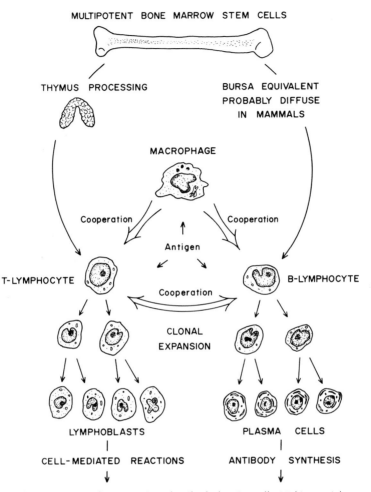

MULTIPOTENT BONE MARROW STEM CELLS

THYMUS PROCESSING

BURSA EQUIVALENT
PROBABLY DIFFUSE
IN MAMMALS

MACROPHAGE

Cooperation

Cooperation

Antigen

T-LYMPHOCYTE

Cooperation

B-LYMPHOCYTE

CLONAL
EXPANSION

LYMPHOBLASTS

PLASMA CELLS

CELL-MEDIATED REACTIONS

ANTIBODY SYNTHESIS

Fig. 9.1.   Major steps in the maturation of antibody-forming cells. Multipotential stem cells:
located in the bone marrow, are transported either to the thymus (in the case of T-cell development) or through the "bursa equivalent," probably a diffuse network of lymphoid organs (in
the case of B-cell development). When presented with antigen, the T lymphocytes and B
lymphocytes cooperate with the macrophage as indicated and undergo a cellular transformation. This results in clonal expansion and eventual formation of mature lymphoblasts or of
plasma cells, which are capable of cell-mediated reactions and of antibody synthesis,
respectively.

antibody (Fig. 9.1). According to the clonal selection theory, each lymphocyte is
committed to the production of a clone of cells, all of which produce one and
only one antibody species (Burnett, 1959). Three hypotheses were developed to
account for antibody diversity:

1. *The Instructive Hypothesis.* Any lymphocyte is capable of reacting to any antigen by producing an antibody molecule that can mold itself to the shape of the antigen.

2. *The Germ Line Hypothesis.* One gene exists for every possible antigen. This hypothesis proposed that these genes arose in the course of evolution and that a substantial part of the genome of higher organisms must be given over to a repository of information that may never in the course of the life of the individual organism be put to use.

3. *The Somatic Mutation Hypothesis.* The chromosomal region carrying the genes for antibodies is particularly unstable and undergoes mutation at a very high rate. Through a process of hypervariability, new gene sequences are constantly being generated, supplying each individual with a large number of antibody molecules. This hypothesis would imply (1) that an individual lymphocyte is capable of producing only one particular antibody species and (2) that the mutation rate for the immunoglobulin loci is extremely high.

The instructive hypothesis was eliminated many years ago. When antibody molecules are denatured by treatment with urea they will readily reassociate; they retain the original specificity for the antigen. This results from the re-establishment of their tertiary structure, which must be dependent upon the primary amino acid sequence of the antibody molecule. Therefore, any particular antibody must be coded for by a particular gene which specifies the primary amino acid sequence of the antibody. More direct evidence bearing on this point comes from work with single plasma cells. The fact that such cells respond to only a single or closely related antigen is strong evidence against the instructive hypothesis. Recent evidence suggests that elements of both the germ line and somatic mutation hypotheses may be correct.

III. THE STRUCTURE OF THE ANTIBODY MOLECULE

The antibody molecule is Y-shaped and consists of a constant region and a variable region which determines the particular specificity for the antigen (Dreyer and Bennett, 1965). There exist four chains in all: two light (L) and two heavy (H). The four chains are held together by disulfide bonds. Thus, two light chains possess variable and constant portions, as do the two heavy chains (Fig. 9.2). Each variable region in turn is divisible into four regions (referred to as framework regions) which do not vary greatly from one antibody molecule to another, and three hypervariable regions which are greatly different in different antibody molecules and are folded to form the antigen-binding region (Robertson, 1980). The light chains are of two types: kappa (κ) and lambda (λ). The heavy chains

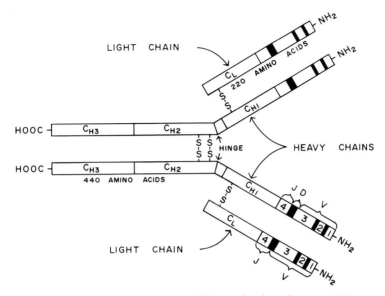

Fig. 9.2.   Diagram of the immunoglobulin molecule (Robertson, 1980).

constitute eight types, all of which are determined by constant region differences. The heavy chains (μ, γ_1, γ_{2a}, γ_{2b}, γ_3, γ_4, δ, ϵ) define the immunoglobulin classes (IgM, IgG$_1$, etc.) (Gally, 1973). The N-terminal sequence is in the variable portion of the molecule. The term "constant region" is somewhat of a misnomer and applies only in a relative sense. This is most strikingly demonstrated by the existence of "allotypes." Allotypes are genetic variants of the antibody molecule which differ among different individuals; allotypic variation occurs in the constant portion of the immunoglobulin gene.

IV. THE ARRANGEMENT OF ANTIBODY GENES

The genes responsible for specifying the structure of the antibody are numerous and diverse, and their number and arrangement on the chromosome is the subject of substantial research. For light chains it has been shown that three different DNA segments code for the constant and variable portions of the antibody molecule (Seidman and Leder, 1978): the constant (*C*), variable (*V*), and *J* genes. These genes are widely separated in the embryo and become rearranged through a splicing mechanism prior to the formation of the mature antibody-producing cell. After the rearrangement they lie close to one another (Fig. 9.3) and yield a κ or a λ immunoglobulin light chain. This process requires bringing together a DNA segment coding for one of multiple variable regions with a DNA segment coding for the other appropriate constant region. The λ and κ chains are determined by two different gene families. In the mouse there are two copies of the λ gene and between 100 and 300 copies of the κ gene (Robertson, 1981). Thus, the formation of a functional, mature light chain gene involves first the selection of a λ or κ gene and its juxtaposition next to one of four possible J genes. This most likely occurs by deletion of all the intervening DNA between these elements (Cory and Adams, 1980). The J gene codes for a stretch of 11 amino acids and lies between the variable and constant regions; however, the J gene is separated from the C gene by an intervening sequence (Gough et al., 1980). The function of the J region may be to provide a recombination point for the V DNA and may also itself contribute to antibody diversity (Schilling et al., 1980). The final light chain gene will possess a large intervening sequence in the DNA that will be transcribed and subsequently removed from the messenger during processing.

For heavy chains a similar situation prevails, although it is somewhat more complex because of the many different classes of constant heavy chains and because of another variable sequence, the D gene (Fig. 9.2). Hybridization studies have shown that there are at least 50 V_H genes, and probably many more, arranged in the chromosome of embryonic cells in tandem arrays and widely separated from the C genes (Cory and Adams, 1980). There appear to be approximately 15 copies of the C region per genome and a much larger number of V region genes (50 to 1000), although this is a more problematic estimate because of its extreme variability. The V and J genes (of which five copies exist) are joined by the D region. This D (for "diversity") region may undergo considerable variability in length and also codes for a hypervariable region (Robertson, 1980). Assembly of an operational heavy chain gene proceeds by selection of one of the many variable genes followed by a D and then a J gene, so that the three are spliced contiguously to one another. This complex will still be separated from the appropriate constant gene by a large intervening segment of DNA; these heavy chains will be transcribed and processed as are the light chains (Fig. 9.3).

The heavy chains have an additional modification: "the switch." In the course of the antibody response, the first class of antibodies to be produced is IgM, followed by IgG and, later, by IgA and IgE. Because each of these molecules has a different C_H portion but an identical V_H region, a mechanism for making the proper assembly must exist. Thus, the same V_H sequence will be expressed first in μ chains and later as a part of a γ, α, or ϵ chain. Although the assembly of the variable chains involves selection and splicing of different segments, the heavy chain switch requires actual deletion of other constant region genes (Molgaard, 1980). Studies of hybridization kinetics on different plasmacytoma cells establish that the deletion pattern is consistent with a C_H gene order $\mu-\gamma_3-\gamma_1-\gamma_{2b}-\gamma_{2a}-\epsilon-\alpha$. An exception to this pattern is IgD, an immunoglobulin that is

EMBRYONIC   DNA   UNDIFFERENTIATED

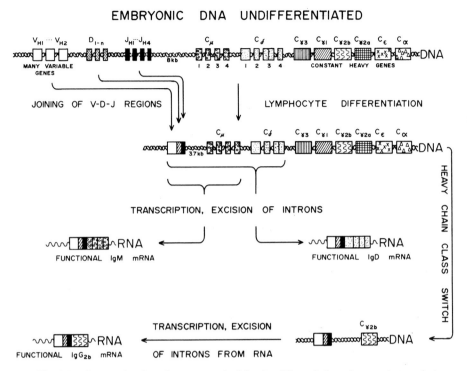

Fig. 9.3. Proposed series of events required for the differentiation of mouse heavy chain genes. The light chain immunoglobulin genes are believed to be assembled in an analogous manner, although switching is not required.

expressed simultaneously with IgM. In this case it appears that the $\mu-\lambda$ genes, which are located between the V_H gene and the δ genes, are simply treated as an intron. The whole region $V_H-\mu-\delta$ is transcribed and then the intervening μ segment is excised in the processing of a functional IgD molecule (Robertson and Hobart, 1981; see Fig. 9.3).

There is another unique feature of the immunoglobulin system that should be mentioned. This is the expression of only a single allele (paternal or maternal) of each antibody gene in any given cell. Although reminiscent of X chromosomal inactivation (see Chapter 13), the basis for this phenomenon is almost certainly different. Coleclough *et al.* (1981) argue that nonproductive rearrangements of the nonfunctioning allele are responsible, at least in part, for allelic exclusion. This could result from (1) lack of a proper RNA processing signal, (2) formation of translational stop codons or other errors that extinguish transcription, or (3) selection of nonfunctional DNA segments.

The discovery that immunoglobulin molecules are determined by several genes whose products are assembled together seemed to violate the "one gene–one polypeptide" hypothesis. In fact, it had been known for a long time from studies on hemoglobin and other molecules that a final gene product could consist of a number of polypeptides that are determined by different genes. However, the observations on the determination of antibody structure seemed to argue that two or more genes might act together to produce a single polypeptide. If this were correct, it would represent an extremely novel finding and would require molecular mechanisms for the fusing of the genes or their products.

Some of the strongest evidence in favor of the multiple genes–one polypeptide model for production of the variable and constant regions comes from studies by Fudenberg on a patient with biclonal gammopathy. This is a disorder occurring in humans in which two serum proteins are found in abnormally elevated levels. It is most commonly characterized by elevated IgG and IgM, and the disease may be either benign or malignant. The fact that this and related disorders arise from the transformation of single clones of plasma cells has been extremely useful in the analysis of immunological phenomena. The particular individual with this condition demonstrated an elevated immunoglobulin level of two different antibody molecules, IgG_2 and IgM. Both these proteins were sequenced, and it was found that the variable portions possessed identical amino acid sequences, whereas the constant regions for the two immunoglobulin types contained different amino acids in 65% of the positions. The C_γ and C_μ regions are, of course, known to be determined by different genes, and so it is to be expected that there would be many differences in their amino acid sequences. The identity of the variable regions is striking and deserves comment. The C_γ and the C_μ genes are quite different and possess many amino acid substitutions resulting, no doubt, from their wide evolutionary divergence (Fig. 9.4). The fact that the V regions in the IgG and IgM molecules of this patient are identical would thus constitute strong evidence that the V and C regions of the immunoglobulin molecule are determined by different genes and that their connection in the final protein product results from subsequent events involving the processing of the genes, messengers, or final proteins.

Furthermore, the fact that the IgG and IgM variable portions are identical indicates that no mutations occurred in this region after the point in embryonic development at which the cell types that would give rise to IgG- and IgM-forming cells in the adult diverged. Therefore, if the somatic mutation hypothesis is correct and if antibody diversity is generated through recombination and mutation, antibody diversity cannot be occurring randomly (see Chapter 1). If such variation *were* occurring randomly, then one would expect that mutations would have occurred in the variable portion of the molecule in the periods both before and after the time of divergence of the cell populations in the embryo, and the IgG and IgM molecules would possess at least a few differences in their variable

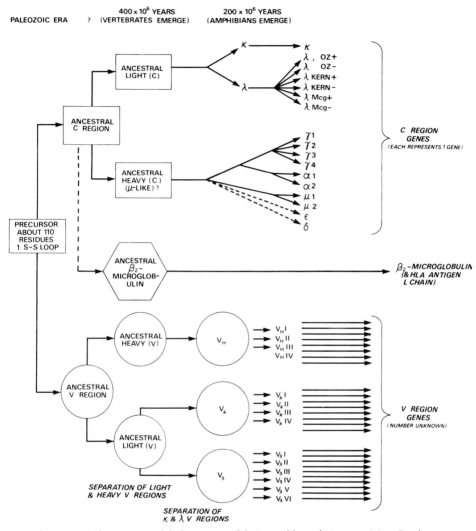

Fig. 9.4. Modern immunoglobulin genes and their possible evolutionary origins (Goodman and Wang, 1978).

regions. Thus, the somatic mutation theory in its simplest sense cannot be correct.

Much of what we know of the origin of antibody diversity has come through the use of myeloma tumors, in particular the mouse plasmacytoma (MPC) and mineral oil plasmacytoma (MOPC) cell lines. Such tumors may be either spontaneous or induced and are capable of producing large quantities of antibody of unknown specificity. Non-immunoglobulin-producing mouse myeloma cells can be easily isolated by cloning in soft agar overlaid with antisera against mouse κ chains. Clones that are positive for antibody production possess a "halo" that is caused by precipitation zones between the immunoglobulin and the antiimmunoglobulin antibody. Such investigations have shown that the mutation rate to nonproduction (as measured by the fluctuation test) is approximately 1.1×10^{-3} per cell per generation (Coffino and Scharff, 1971). Milstein isolated and characterized four mutations occurring in the variable portion of the γ chains of the MOPC-21 cell line. He found that each could be identified as either a missense or a deletion mutation. Furthermore, the fact that the mutations appear to be relatively localized argues for preferential sites of mutation and suggests that these mechanisms in myeloma cells are the same by which antibody diversity *in vivo* is generated.

Cotton and Milstein (1973) has also shown in somatic cell hybrids derived from fusion of cells producing antibodies of different specificity that heteromorphic immunoglobulins are not produced. This fact implies that the variable–constant gene integration is preserved even though heavy–light chain hybrid molecules are generated.

The basis, then, of antibody diversity lies in several mechanisms combining features of both the somatic mutation and the germ line hypotheses: (1) Large numbers of antibody genes exist in embryonic cells. (2) The *V*, *J*, and *D* genes can recombine at random and therefore can generate many different combinations. (3) The point of connection of the *V* and *J* genes is not always exactly the same, causing additional variability in the region of amino acid 96 which is critical for binding to the antigen (Marx, 1981). (4) Light and heavy chains can combine at random, providing another major source of variability. (5) Actual amino acid substitutions occur in the hypervariable regions, particularly regions *1* and *2*. With all these mechanisms available, the ability of the immune system to generate an apparently unlimited range of different antibody molecules is not surprising.

V. IMMUNOGLOBULIN EXPRESSION IN CELL HYBRIDS

Much of the previous discussion serves as a framework to place into perspective the analysis of immunoglobulin expression in cell hybrids. By using cell fusion, it is possible to put different immunoglobulin producers (and nonproducers) in the same cytoplasm in order to analyze their molecular control mechanisms. Several types of findings indicate that alternate types of regulation are occurring simultaneously (Table 9.1). If an immunoglobulin-producing cell is hybridized with another producer, both parental types of immunoglobulin will

TABLE 9.1 Expression of Immunoglobulin Production in Somatic Cell Hybrids between Producing and Nonproducing Cell Lines

Cell lines	Cell type	Animal of origin	Ig produced	Ig produced in hybrids	Assay method	References
MOPC-315	Plasmacytoma	Mouse	Iga	$\pm$ Ig	Antigen binding by labeled Ig	Periman (1970)
C1 1D	Fibroblast-like	Mouse	None			
MPC-11	Plasmacytoma	Mouse	IgG_{2b}, kappa	None	Immunoprecipitate of labeled Ig	Coffino et al. (1971)
	Fibroblast-like	Mouse	None			
T5-1	Lymphoblast	Human	Lambda	Human lambda	Immunofluorescence	Orkin et al. (1973)
3T3C2	Fibroblast-like	Mouse	None			
Thymocytes		Human	None	Human IgG	Immunoprecipitate of labeled Ig	Parkman et al. (1973)
Fibroblasts		Mouse	None			
266B1	Myeloma	Human	IgE, lambda		Immunofluorescence	Zeuthen et al. (1976)
L929	Fibroblast-like	Mouse	None	None	Immunofluorescence	
HeLa	Epithelial-like	Human	None	None	Immunofluorescence	
Fibroblast		Human	None	None	Immunofluorescence	
Glia-like		Human	None	None	Immunofluorescence	
MPC-11	Plasmacytoma	Mouse	IgG_{2b}, kappa	None	Immunofluorescence	Maurer and Morrow (1977)
LMTK	Fibroblast-like	Mouse	None			

be produced in the hybrid. In some cases immunoglobulin synthesis can be activated in interspecific hybrids in which a producer is fused with a cell (for instance, a human thymocyte) that may possess the potential for forming immunoglobulin. However, fusion of a fibroblast with a myeloma cell produces a very different result: no immunoglobulin is produced. Indeed, even fusion of an overproducing polyploid MPC cell with a mouse L cell fails to allow immunoglobulin synthesis (Maurer *et al.*, 1982). This absence of a dosage effect in polyploid hybrids is in striking opposition to the results that have been culled in the other cell hybrids between differentiated and undifferentiated cells (e.g., melanoma and liver cells; see Chapters 7 and 8).

It therefore appears that the immunoglobulin loci represent genetic regions with a high degree of instability but are confined in this variability by a certain degree of both developmental and spatial restriction. The antibody molecule is a joint effort resulting from the fusion of genes originally located some distance from one another and rearranged in the course of embryonic development. Even when the variable and constant region genes come into close proximity with each other, there still exists between them an untranslated segment of approximately 1000 nucleotides. The instability of the immunoglobulin loci observed in cultured myeloma cells is a reflection of the mechanism by which antibody diversity is generated.

Antibody synthesis appears to be controlled by at least two different mechanisms:

1. "Cis-recessive" control, occurring in hybrids between cells derived from the reticuloendothelial system. If both cell types are capable of producing immunoglobulin, production of both types of immunoglobulin molecules will be continued in the hybrid. A previously inactive cell type will be activated (i.e., human thymocyte × mouse myeloma), also resulting in the production of both types of immunoglobulin (human and mouse) in the hybrid (Schwaber and Cohen, 1973, 1974).

2. A second order of regulation concerns hybrids between non-immunoglobulin-producing cell types (i.e., mouse fibroblasts) and producers. Immunoglobulin production in such hybrids is extinguished (Coffino *et al.*, 1971), and attempts to titrate out a hypothetical repressor from the fibroblast parent by increasing the dosage of myeloma genomes have been unsuccessful (Maurer *et al.*, 1982). This "trans-dominant" control appears to be different from the former type and may reflect a different mechanism of control.

VI. MONOCLONAL ANTIBODIES PRODUCED BY CELL HYBRIDS

The fact that cell hybrids between related reticuloendothelial elements continue to produce immunoglobulin suggested a means by which monospecific antibody species might be produced in culture. Such a system was developed by fusing spleen cells from immunologically primed animals with thioguanine-resistant myeloma cell variants (Köhler and Milstein, 1975, 1976; Fig. 9.5). The resulting hybrids (referred to as "hybridomas") are selected with HAT medium (Chapter 2), which kills off the myeloma parent. After several weeks clones that will produce monoclonal antibodies arise, and in many cases clones producing immunoglobulins specific to the antigens used to immunize the host can be isolated. The production of monoclonal antibodies has been the subject of a thorough investigation by Fazekas de St. Groth and Scheidegger (1980). They have isolated a variant myeloma clone that has a generation time of 8.7 hours, fuses effectively with B lymphoblasts, and produces no immunoglobulin chains

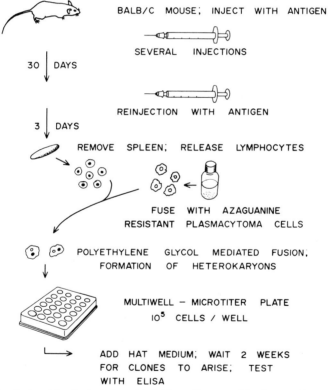

Fig. 9.5. Protocol for the isolation of hybridoma lines capable of generating monoclonal antibodies. "ELISA" (enzyme-linked immunosorbent assay) refers to one of many procedures available for the detection of those hybridomas capable of producing the particular antibody desired. This particular method uses an antibody to mouse immunoglobulin that is linked to an enzyme such as alkaline phosphatase or peroxidase. Enzyme activity is used as the criterion of antibody production by a particular clone.

of its own. Their investigation of the variables involved in the production of hybridomas has enabled them to standardize the procedures, resulting in high and consistent yields of clones producing monoclonal antibodies. Their principal findings are the following: (1) that the addition of peritoneal macrophages to the mixtures will greatly increase the yield of hybrid clones (apparently by cleaning up detritus); (2) that reagent-pure polyethylene glycol of 4000 MW (autoclaved) is the most effective fusogen; and (3) that the reverse selective HAT medium could be added immediately after fusion, thereby expediting the procedure.

Although a number of questions have yet to be resolved, including the stability and ability of such hybrids to continuously produce the antibody in question, the system promises to be a significant one for the resolution of a number of important questions (Table 9.2). Probably the most important use of monoclonal

TABLE 9.2   Examples of Basic and Clinical Problems Approached through Use of Monoclonal Antibodies

Question investigated	Antibodies produced	Reference
Possible autoimmune basis of myasthenia gravis	Anti-nicotinic acetylcholine receptors	Lennon and Lambert (1980)
Function of the major histocompatibility complex (MHC)	Alloantibodies against rat MHC antigens	McKearn et al. (1979)
Vaccination against malaria	Antibodies against sporozoite surface antigens	Cox (1980)
Standardized sera for tissue typing	Anti-human HLA-A, -B, -C, and -D antibodies	Milstein et al. (1977)
Analysis of antigenic determinants on malignant cells	Antibodies against human melanomas	Koprowski et al. (1978)
Antigenic analysis of viral determinants	Anti-influenza	Koprowski et al. (1978)
Analysis of thymocyte differentiation antigens	Anti-rat thymocyte membranes	Williams et al. (1977)
Analysis of human leukemia antigen	Anti-ALL (acute lymphoblastic leukemia)	Ritz et al. (1980)
Antileukemia immunotherapy	Anti-thymus differentiation antigen	Bernstein et al. (1980)
Interstrain variation of rabies virus coat	Anti-rabies	Wiktor and Koprowski (1978)
Location of genes coding for human cell surface antigens	Membranes on human T lymphocytes	Barnstable et al. (1978)
Identification of different photoreceptors	Adult rat retina membranes	Barnstable (1980)

antibodies is one in which a complex group of antigens are injected into the host and a panel of hybrid clones are screened until one obtains that clone that produces the desired antibody. This approach can avoid long and arduous purification of the antigen or difficult immunoadsorption techniques with the γ-globulin fraction to purify the antibody taken from the immunized animal.

An even more interesting feature has been pointed out by Springer *et al.* (1978). When cells are used as the immunogen, the host response is particularly complex because the injected cells carry an extremely intricate array of antigenic structures. Through use of the monoclonal antibody system it is possible to isolate lines that produce antibodies to single surface antigens. By using this approach investigators produced 10 different rat monoclonal antibodies to mouse antigens. These antibodies fell into four groups, each defining determinants that possessed particular specificities for different cell types. The results so far obtained strongly suggest that differentiated functions exist and appear on certain cell types at certain times and that the hybridoma system is a profitable way to investigate them.

Another area of active investigation uses the hybridoma to construct monoclonal antibodies against commonly infectious and clinically important microorganisms. These techniques detect subtle differences in viruses or bacteria not detectable by conventional methods. This method has been employed (Kaplan and Koprowski, 1980) to demonstrate different substrains of rabies virus.

Other applications of such antibodies include detection of microbial antigens in tissue specimens and (in the future) the development of specific antibodies for use in the treatment of bacterial, protozoan, and viral diseases. This latter application requires the development of a human hybridoma, because any such therapy would necessarily employ human immunoglobulin.

VII. CONCLUSION

Our understanding of the immune system has grown from the sketchy framework of 20 years ago into a detailed picture that now proposes many specific molecular explanations for the manner in which antibodies are generated. Of the three hypotheses designed to explain the means by which clonal selection is brought about, features of both the germ line and somatic mutation models are satisfactory. The work of Cory and Adams (1980) suggests substantial germ line diversity: the number of V_H genes may reach into the hundreds. On the other hand, the somatic mutation hypothesis still has features that appear to be correct (Bernard *et al.*, 1978), and a resolution of this question will require a synthesis of the somatic mutation and germ line theories. Because a functional Ig gene is formed by deletion of all of the intervening DNA between the *V, J,* and *D* genes (Davis *et al.*, 1980), resulting from the splicing of different DNA molecules, the somatic recombination model of diversity is an appealing one. However, even though there is recognition that in theory different immunoglobulin genes might provide sufficient diversity for antibody structure, in practice it is currently believed that further diversity through mutation is required (Robertson, 1980).

The studies on cell hybrids have yet to provide a satisfactory molecular explanation for the regulation of the immunoglobulin genes, but they indicate that at least two levels of regulation are occurring. A most valuable side result of these investigations is the development of hybridoma technology, which takes advantage of the ability of myeloma–spleen cell hybrids to produce immunoglobulin.

10

Hypotheses of Malignancy and Their Analysis through the Use of Somatic Cell Hybrids

I. INTRODUCTION

Cancer biology is an immense and highly diversified area of research with a history going back into the nineteenth century (Boveri, 1920). In 'this chapter we will consider a broad portrait of our theoretical understanding of neoplastic transformation. We will then discuss in some detail experimental systems—in particular, cell hybrids—that have been employed in its analysis. As will become evident, much of the evidence is contradictory, and a resolution of conflicting lines of investigation lies some distance in the future. Even so, broad outlines of a model of malignant change are developing, and many specific predictions are under experimental analysis.

There are different phases of unrestricted growth, and they are defined operationally. Thus, a "malignant," "cancerous," or "neoplastic" cell is one that, when injected into a histocompatible individual, will cause the formation of a tumor that grows progressively and kills its host. Such cells may differ tremendously in their degree of malignancy and invasiveness; in some cases, as few as one cell is sufficient to initiate formation of a tumor. A cell may also take on some properties of a cancer cell without acquiring the trait of malignancy. Such cells are transformed and have assumed the characters of unrestricted growth, failure to experience contact inhibition, growth in soft agar, and karyotypic abnormalities. Thus, transformation is necessary but not sufficient for tumor genesis, and these two properties can be separated in somatic cell hybrids (Stanbridge and Wilkinson, 1978).

The most widely supported model today is that the change from the normal to a malignant state involves a series of mutational alterations that occur in the host cell DNA and that introduce an increasing liability for unrestricted growth. The term "mutational" is used here in a broad sense to include all heritable changes within the cell nucleus, including transposition of genetic material (Cairns, 1981). This derepressed or uncontrolled growth pattern, although fundamentally genetic, can be precipitated by environmental stimuli or through the introduction of genetic information by oncogenic viruses. The relative weight of these factors, their role in particular types of cancers, and their patterns of interactions or synergism are complex and remain to be unraveled.

II. MODELS OF CANCER

Overwhelming evidence exists for a strong and perhaps decisive genetic component in the etiology of malignancy. This evidence has been thoroughly reviewed by Ames and his collaborators (McCann and Ames, 1976) and is most convincingly supported by his own comprehensive studies in which an excellent correlation between mutagenicity and carcinogenicity for a vast number of chemical substances has been established (Table 10.1). This information has been attained through the development of a screening procedure that tests substances for their mutagenicity.

Theoretically, it should be possible to test suspected carcinogens for potential mutagenicity simply by measuring their ability to induce reversions to prototrophy in bacterial strains that are auxotrophic for amino acids or other metabolites. This is ordinarily accomplished by plating cells on minimal medium after mutagenization and counting the numbers of revertants. In fact, such studies gave ambiguous and contradictory results, mainly because of technical problems, which can be resolved by the "Ames Test." This protocol uses tester strains of bacteria (*Salmonella*) into which the following modifications have been introduced:

1. Several histidine-requiring mutations including both missense and frame shifts. This takes account of the fact that mutations can be induced by a number of molecular mechanisms.

2. A mutation (*rfa*) that affects the lipopolysaccharide coat and that expedites the entry of fat-soluble substances into the cell.

3. A *rec⁻* mutation that causes a deficiency for repair of uv-induced mutational damage (*uvrB*).

4. The addition of a liver microsome supernatant fraction that metabolizes noncarcinogenic compounds to carcinogenic compounds.

When such multiply marked tester strains are used to screen possible carcinogenic compounds, the excellent correlation shown in Table 10.1 is obtained.

TABLE 10.1   Summary of the Relationship between Carcinogenicity and Mutagenicity for a Large Number of Compounds[a]

Test compound	Nonmutagenic	Mutagenic
Noncarcinogens	A = 94/108 (87%)	B = 14/108 (13%)
Carcinogens	C = 18/175 (10%)	D = 157/175 (90%)

[a] Of the 18 carcinogens that were not mutagenic, many are mutagenic in other systems or produce mutagenic metabolites (McCann and Ames, 1976).

There are, however, a number of exceptions (B and C in Table 10.1) that require further consideration.

Another test system has added to our understanding of the relationship between mutational damage and cancer and has clarified the basis for some of the discrepancies uncovered in the Ames system. Devoret (1979) has reviewed the "inductest" results; in this system a special lysogenic strain of *Escherichia coli* is used to test compounds for their efficacy in the induction of active (lytic) phage. He concludes that carcinogenesis may be the result of indirect damage to the DNA rather than direct mutational action on the part of the test compound. Thus, all carcinogens must be mutagenic, but some mutagenic substances may alter the host cell DNA without causing sufficient damage to the hereditary material to result in malignant transformation. In this view, carcinogenesis is an event somewhat akin to virus induction in bacteria and involves an initial blockage of DNA synthesis due to deformation of the DNA molecule by the carcinogen. This brings about the induction of proteins that affect the replicative capacity of the polymerase complex and are able to induce both secondary mutational damage and production of active virus particles.

A number of other types of studies from a variety of disciplines point to somatic mutation as a mechanism in carcinogenesis.

1. Many tumor types may have a clonal origin. Leiomyomas of the myometrium arise from single cells, as shown by an analysis of X-linked gene products (Chapter 13). Myeloma tumors consist of cells, all of which produce identical variable chains; this finding indicates that each tumor arose from a single original plasma cell (Chapter 9).

2. Some cancers show a simple Mendelian pattern of inheritance. Hereditary retinoblastoma is an example of such a cancer; it is inherited as a dominant gene with approximately 80% penetrance. Knudson *et al.* (1975) have proposed that in such families a cancer-causing gene is segregating in the pedigree and that a second somatic mutation is required for expression of the tumor. Comings (1973) has expanded this model by postulating that mutation of both alleles in a recessive cancer regulatory gene could explain the pattern of inheritance observed in neuroblastoma.

3. In cases in which an environmental component is indisputable, genetic factors appear to be responsible for a strong predisposition to cancer. The fact that only a certain fraction of heavy cigarette smokers contract lung cancer argues for the existence of certain genetically determined responses such as hydrocarbon metabolism, DNA repair, and immunological surveillance (Vogel and Motulsky, 1980).

On the other hand, a vast body of evidence that has been collected over the years demonstrates that at least some cancers in experimental animals have a clear and demonstrable viral etiology. The role of viruses is most convincingly demonstrated through the use of temperature-sensitive mutants of both RNA and DNA tumor viruses. Such mutants will cause malignant transformation at the permissive but not at the restrictive temperature, and the cell phenotype can be reversed from transformed to normal depending on whether the cultures are maintained at the permissive or the restrictive condition. The reversibility of the phenomenon demonstrates the persistence of the viral genome (see review by Ozer and Jha, 1976).

Perhaps the strongest candidate for a direct viral etiology in the development of human malignancy is the Epstein–Barr (EB) virus, which appears to be responsible for Burkitt's lymphoma and nasopharyngeal carcinoma (Henle *et al.*, 1979). The following lines of evidence support a case for a causal relationship between this malignancy and the virus: (1) the detection of viral DNA or viral antigens in the tumor; (2) the transformation of cultured B lymphocytes by the virus; (3) the induction of lymphomas by the virus in nonhuman primates; (4) the demonstration of higher antibody titers to the virus in lymphoma patients compared with the titers in a control population; and (5) the correlation between antibody titer patterns and the prognosis of the tumor. The EB virus is definitely the cause of infectious mononucleosis, which may represent a benign, self-limiting lymphoma. One of the striking properties of malignant cells is the presence of a small reciprocal translocation involving chromosomes 8 and 14. The presence of chromosomal changes in malignant cells is not uncommon and may have fundamental implications (see later). However, it is also clear that other changes are required in the cell or its environment in order for a malignancy to develop in an infected individual.

Recent investigations utilizing recombinant DNA technology have provided insight into the mechanism by which viruses produce oncogenic changes. Oncogenic viruses, such as retroviruses, possess RNA sequences homologous to cellular genes from vertebrates, and there is reason to believe that these viral genes originate from cellular DNA referred to as "protooncogenes" (Bishop, 1981). These genes may have a regulatory function under normal circumstances and may have become detached in the course of evolution. Their reintroduction into the cell may cause a regulatory disturbance because of the extra dose and may result in malignancy. Alternatively, in a mutated condition they may introduce altered (and, therefore, carcinogenic) gene products into the cell.

Because of the relative simplicity of such viruses, it is possible to identify those oncogenes and study their functions by reintroducing them into untransformed cells. Recent experiments using transfection techniques provide an assay for transformation of mammalian cells and have shown that transformation can be affected using sequences either extracted from already-transformed cells or from oncogenic viruses. One such oncogene has been studied in particular detail, and this is the *src* gene from Rous sarcoma virus. It produces a protein known as pp60V-*src,* which is a phosphoprotein of 60,000 daltons which functions as a protein kinase specifically phosphorylating the amino acid tyrosine. It is assumed that the modification through phosphorylation of a number of cellular proteins would have profound effects upon cellular metabolism, thus resulting in the transformed phenotype. This model would support the idea that malignancy would be a drastic but stable alteration in the cellular phenotype.

An alternative view that considers cancer as a particular state of differentiation has-been urged by the innovative studies of Mintz and Illmensee (1975). Teratocarcinoma cells are malignant stem cells derived originally from gonadal tissue, either spontaneously or by grafting early embryos into testis. Ordinarily, however, they contain carcinoma cells which are highly malignant. These teratocarcinoma cells can be maintained as ascites tumors known as embryoid bodies. The embryoid body possesses an outer "rind" of abnormal cells and an

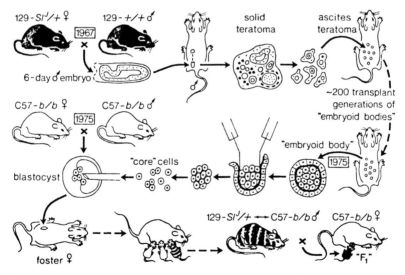

Fig. 10.1. Protocol for the production of mosaic mice from teratocarcinoma cells. (From Mintz and Illmensee, 1975.)

interior composed of cells with ostensibly normal karyotype; and the development of a chimeric animal can be initiated by injection of such ''core'' cells into a blastocyst of a differing genetic strain (Fig. 10.1). In some cases animals can be produced in which the genetic contribution of both the embryo and the tumor cell can be recognized. In fact, some mosaic combinations can give rise to animals in which the genotype of the teratocarcinoma cell is represented in every adult tissue (Fig. 10.2). Thus, a cell type that is highly malignant in one environment will behave completely normally in another. These findings, in the opinion of the authors, ''furnish an unequivocal example in animals of a non-mutational [*sic*] basis for transformation to malignancy and of reversal to normalcy.'' Harris (1979) has, however, reviewed the evidence for epigenetic mechanisms in the etiology of malignancy and feels that it is ''very weak.'' He mentions studies on

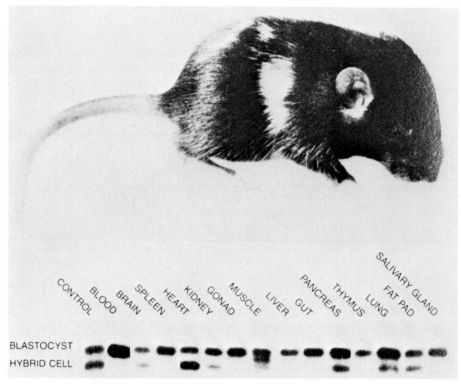

Fig. 10.2. Isozyme patterns and phenotypic appearance of a mouse produced from the fusion of a mouse–human cell hybrid with a teratocarcinoma cell. The slower moving hybrid cell enzyme form of glucosephosphate isomerase is present in seven tissues. (From Illmensee and Stevens, 1979.)

plant tumors in which the malignant phenotype is held in abeyance and suggests that the same situation may prevail in the teratocarcinoma system.

III. ANALYSIS OF MALIGNANCY IN INTRASPECIFIC SOMATIC CELL HYBRIDS

The notion that malignancy could be analyzed by fusing normal and malignant cells occurred to workers at the time cell hybridization was discovered. The early studies of Barski *et al.* (1960) suggested that malignancy behaved as a dominant character in somatic cell hybrids. Although the hybrid product of two lines of differing carcinogenic potential was highly malignant, subsequent studies have indicated that the problem is more complex. At the time of the discovery of cell hybridization, experiments were carried out in which intraspecific hybrids were made between high and low tumor-producing lines. Such investigations resulted in hybrid lines that inherited the phenotype of the more malignant parental cell type (Barski and Cornefert, 1962; Scaletta and Ephrussi, 1965; Ephrussi *et al.*, 1969). Thus, it was presumed that the property of malignant growth was dominant to the inability to form invasive tumors. However, the chromosomes of such hybrid cells were not examined in detail, and subsequent investigations by Harris, Klein, and their co-workers indicated that malignancy was recessive and that the loss of chromosomes in hybrid cells resulted in the exposure of malignancy-producing genetic factors (Harris and Klein, 1969).

Harris and his collaborators have continued this extensive series of investigations over the last decade (1970–1980). Among his earlier studies (H. Harris, 1971) were investigations hybridizing Ehrlich ascites cells and normal diploid fibroblasts obtained from 16-day-old embryos. Such hybrid clones are very unstable and lose chromosomes rapidly and are, therefore, rather uninformative. Another disadvantage of this sytem is the fact that mouse embryo cells transform rapidly in culture and are not an ideal system. A second series of hybrids were analyzed including (1) SEWA (a highly malignant near-diploid tumor) × mouse embryo cells and (2) TA3 (mammary adenocarcinoma) × mouse fibroblasts. Those clones with substantial chromosome loss showed 90–100% tumor incidence, whereas those with an intact chromosome set had a greatly reduced incidence of tumors. In all cases malignancy is much reduced provided a complete chromosome set is retained. Harris claims that specific chromosomes carry factors that suppress malignancy and that these factors are lost in the chromosomally reduced hybrid. The failure of the tumors to grow because of histoincompatibility is not a factor since the results are the same with allogeneic irradiated animals; thus, the conclusion is that malignancy behaves as a recessive character.

In another series of experiments it was observed that mouse A9 fibroblasts,

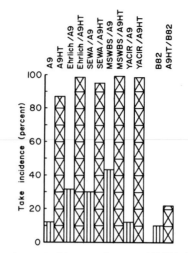

Fig. 10.3. Cumulative tumor incidences of tumor–A9HT hybrids compared with those previously obtained for tumor–A9 hybrids. The take incidence of hybrids between A9HT cells and B82 cells is also shown. (From Wiener et al., 1973.)

which have extremely low tumorigenic potential, suppress several highly malignant lines in hybrid combinations. The A9 (azaguanine resistant, dies in HAT) cells are transformed but nonmalignant. However, the take incidence with a highly malignant derivative of A9, known as A9HT, was *much* higher (Fig. 10.3). The A9HT line was selected by injection of A9 cells into mice and isolation of rare, tumor-producing variant cells. This further strengthens the belief that malignancy results from specific genetic factors.

Subsequent investigations used a variety of tumor-producing cells hybridized to other malignant types. All crosses produced a high percentage (90–100%) of tumors except for a methylcholanthrene-induced sarcoma (MSWBS) that was hybridized to YACIR, a Maloney virus-induced lymphoma; the percentage of tumors from this hybrid was only reduced to 63%. The fact that such malignant crosses fail to complement one another suggests that a single "locus" is responsible for malignancy. However, Jami and Ritz (1973, 1975) have hybridized malignant L cells with highly malignant leukemia cells, and the hybrid is nonmalignant. The authors argue that the result is *not* due to chromosome loss because the hybrid had a high chromosome number. More recently this group (Aviles et al., 1980) has expanded its investigations and performed extensive chromosomal banding analysis on hybrids derived from fusion of highly tumorigenic LMTK⁻ cells (Chapter 2) and normal mouse thymocytes or fibroblasts. In no case of malignancy was the loss of any specific chromosome pair correlated with tumor formation (Fig. 10.4).

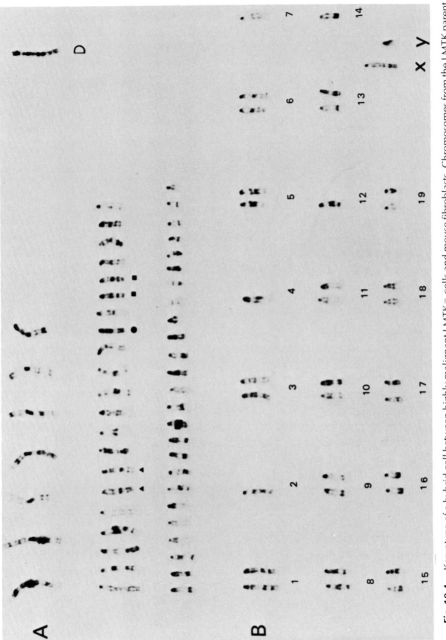

Fig. 10.4. Karyotype of a hybrid cell between highly malignant LMTK⁻ cells and mouse fibroblasts. Chromosomes from the LMTK parent are grouped in A; those from the fibroblast parent in B; D, highly rearranged chromosome containing elements from several normal chromosomes. (From Aviles et al., 1980.)

IV. GENETIC CONTROL OF MALIGNANCY IN
INTERSPECIFIC HYBRIDS

The analysis of malignancy through the use of cell hybrids could be expedited if it were feasible to follow chromosomal segregation in interspecific hybrids. This has become possible by way of several mechanisms that enable testing of tumor formation in xenogenic situations: (1) most importantly, use of the "nude" mouse, an athymic strain that lacks functional T cells and that will accept grafts from different species and even different classes; (2) use of the cheek pouch of the Syrian hamster, an immunologically privileged site; (3) use of embryonated eggs for growth of potentially malignant material. These techniques have opened a new dimension in the examination of malignancy in cell hybrids, making possible the determination of tumorigenicity in human material.

Interspecific hybrids have been studied by Barski *et al.* (1973). These authors fused a BUdR-resistant, mouse, malignant cell line with an azaguanine-resistant cell line; hybrids were malignant by the criterion of tumor formation in the Syrian hamster cheek pouch or growth in embryonated eggs and had undergone extensive chromosomes loss, a result that is in agreement with Harris's observation on intraspecific hybrids.

Kucherlapati and Shin (1979) have analyzed the tumorigenicity of a number of malignant mouse–human diploid fibroblast hybrids. In a total of 63 hybrids tested by injection into nude mice, 62 of these were tumorigenic; combined chromosome and isozyme analyses demonstrated that each of the 24 human chromosomes was represented in some of the tumor-producing hybrids. This would indicate that no single human chromosome carries malignancy-suppressing genes. These results led the authors to the conclusion that (1) the suppression of malignancy observed in cell hybrids is due to cryptic loss of oncogenes or (2) two or more chromosomes acting in concert are necessary to bring about the suppression of malignancy. A different study by Croce *et al.* (1979) also indicated that malignancy behaved as a dominant trait in interspecific somatic cell hybrids. Those hybrids were the result of fusions between HT1080 fibrosarcoma cells and normal human diploid fibroblasts or normal mouse fibroblasts. All the hybrids produced were able to form tumors in nude mice.

The observations of Kucherlapati and Shin (1979) are especially striking in view of similar studies by Jonasson and Harris (1977); these latter studies produced just the opposite results. In hybrids derived from normal human diploid fibroblasts and malignant mouse melanoma cells, these authors observed a suppression of malignancy even after all the human chromosomes were eliminated. Furthermore, hybrids that were prepared by irradiating the human parent and therefore that lacked all human chromosomes also exerted a marked suppression on the incidence of tumor formation in the hybrids. On the basis of this information, the authors argue for an extrachromosomal basis for malignancy and suggest that the centrosome may be the responsible element. Furthermore, Harris (1979) has found that in all cases a particular membrane glycoprotein is correlated with malignancy. This is a modified form of a protein present in normal cells and is characterized by a molecular weight of 100,000 and an ability to bind certain plant lectins; it is believed to be a glucose transport protein (Fig. 10.5). This is a striking observation; however, its significance depends to a large extent on the validity of Harris's recessive model of malignancy.

The dominance–recessiveness controversy could be reconciled if the tumor-producing hybrid cells in the Kucherlapati-Shin (1979) study had lost, through deletion, a fragment of those human chromosomes that supposedly suppress malignancy. Unfortunately, data on the frequency of gene deletion in cell hybrids is not available. However, the number of hybrid cells necessary to obtain a tumor was two orders of magnitude above that required from the highly malignant parent; this would suggest that one hybrid cell per hundred contained a deletion for this region at the time the cells were selected by injection into the host. If we assume that, on the average, 100 generations had passed by the time the cells were tested, we estimate a rate of deletion of 10^{-4} per cell per genera-

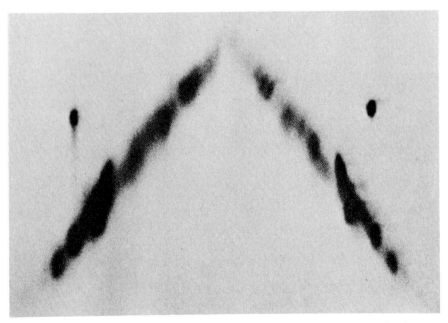

Fig. 10.5. Separation of a membrane glycoprotein whose presence is correlated with malignancy (from Harris, 1979). The figure shows an autoradiogram of gels labeled by affinity binding with ^{125}I-labeled concanavalin A. Separation was achieved by two-dimensional electrophoresis. In the second dimension, dithiothreitol was added to the buffer, causing a depolymerization and separation of aberrant protein from the rest of the membrane components.

tion, an estimate that is quite in line with reality. It would be interesting to know the frequency of deletion of genes such as *HPRT* in cell hybrids and to know whether this figure is compatible with the concept of deletion as a mechanism of tumor production in hybrids.

V. CHROMOSOMAL ALTERATIONS AND MALIGNANCY

Although many observations associating chromosomal changes with malignant transformation have been accumulated over the years, a coherent theoretical framework relating these data has yet to appear.

Hitotsumachi and co-workers (1971) have shown that in virally transformed Syrian hamster cells, the degree of tumorigenicity is related to an increase in the frequency of chromosome 5 and a decrease in chromosome 9. These authors have proposed the existence of "E" (expression) and "S" (suppression) chromosomes, which are assumed to be responsible for the genesis of malignancy. These changes may be the result of position effects, as several other carefully studied cases show a correlation between malignancy and rearrangements.

More recent studies by Azumi and Sachs (1977) also suggest that malignancy may be controlled by an interaction of several genetic factors on different chromosomes. These authors studied mouse myeloid leukemias that failed to respond to a protein known as macrophage and granulocyte inducer (MGI) which causes normal differentiation in most leukemic cells. Such MGI refractory cells possessed characteristic rearrangements involving chromosomes 2 and 12, as shown by Giemsa banding techniques.

These observations, taken with other data, certainly argue for a role of position effects in the process of carcinogenesis. These include the "Philadelphia" chromosome (a translocation from human chromosomes 22 to 9 that occurs in approximately 85% of cases of chronic myelogenous leukemia) and a reversible chromosomal marker associated with malignancy in the mouse (Codish and Paul, 1974). Because normal elements are present in the reverted cell, the abnormal chromosome probably expresses a position effect, i.e., a change in expression due to a transposition of elements.

A mechanism by which such a position effect might be manifested has been suggested by Rowley (1982). She has reviewed the information on consistent chromosomal changes in human malignant leukemias and lymphomas and has pointed out a number of specific chromosomal translocations. It is possible that such translocations result in the juxtaposition of promoter sequences next to genes controlling cellular proliferation, resulting in the activation of patterns of unrestricted growth. The genetic sequences responsible for this transformation may be related to the oncogenes and protooncogenes detectable with DNA probes. Identification and characterization of these sequences is a matter of great interest.

VI. CONCLUSIONS

The alteration from normal to malignant state is accompanied by a number of changes, including a loss of sensitivity to density-dependent growth (Todaro and Green, 1963), growth in semisolid medium, and requirement for serum growth factors (Holley, 1975). Although the question of dominance of malignancy is still in doubt, it is agreed that the transformed phenotype is dominant in hybrids (Ozer and Jha, 1976).

Thus, the sometimes contradictory results reported here are perhaps best understood in the overall framework of our present knowledge of the neoplastic process. Based on studies of human hereditary cancers, cell transformation appears to involve a number of recessive mutational steps. From the studies discussed in this chapter, at least some of these steps may involve chromosomal rearrangements, which could affect the binding of oncogenic viruses to the chromosomes. Croce has provided evidence (1977) that at least two different binding sites for SV40 exist, one on human chromosome 7 and one on 17. The correlation of carcinogenicity with mutagenicity (Ames, 1979) would also be in agreement with a model involving a variety of genetic changes, because mutagenic agents could act either by increasing the frequency of chromosomal aberrations or by causing mutations that facilitate the integration of oncogenic viruses into the chromosome.

Finally, the total suppression of malignancy in malignant teratocarcinoma cells should be mentioned. Although it has been repeatedly argued that malignancy and differentiation could be the result of non-DNA (epigenetic) modifications, Harris (1979) feels that this is extremely unlikely. He suggests that recent findings concerning the changes in DNA sequences that generate antibody diversity (Chapter 9) have dealt a harsh blow to epigenetic hypotheses of cell transformation and differentiation. The dramatic suppression of malignancy in the teratocarcinoma–blastocyst system is somewhat reminiscent of the situation described by Codish and Paul (1974) in which consistent marker chromosome patterns are correlated with tumorigenicity. This suggests a reversible situation in the host cell in which an affinity for oncogenes could be reversed by the presence of the embryonic environment. Such an interpretation would lead to the optimistic belief that malignancy might be reversed through external modifications of the cellular environment (Sachs, 1978).

In summary, it appears that there is no such thing as a single gene for malignancy, but rather that there are mutational alterations of a progressive, multistep nature (Barrett and Ts'o, 1978). This belief is further supported by evidence from the study of hereditary cancers in man (Knudson, 1977) and the demonstration of partial transformation of rodent cells after infection with tumor viruses (Sleigh *et al.*, 1978). In this sense we can think of carcinogens as producing a variety of mutational alterations which are incrementally selected by virtue of their malignancy. It appears likely that the nature of these alterations will be analyzed in the near future on the molecular level. The use of human–rodent cell hybrids in which malignancy is stably suppressed (Sabin, 1981) will be an important tool in this effort, taking advantage of techniques of genetic transfer such as DNA transformation (Chapter 4) combined with recombinant DNA analysis to isolate and thoroughly characterize the responsible genetic entities.

11

Modulation of Gene Expression in Cultured Cells

I. INTRODUCTION

We have already considered differentiation and its control in a number of specialized tissues that are cultivated *in vitro* and that are used as model systems in the study of development. There is, however, another aspect of eukaryotic gene regulation that has been investigated using cell culture systems. In many tissues, variation in the levels of specific gene products can be affected through the use of a wide variety of chemical substances. Some of these compounds, such as hormones and cyclic AMP, are normally encountered by the organism, whereas others have no place in the ordinary existence of the cell (for example, BUdR). In contrast to the process of differentiation, much information is available regarding the molecular mode of action of many of these substances, especially the hormones. An understanding of their role in eukaryotic gene regulation as sophisticated as the bacterial operon model is still some years in the future. However, the efforts of a large number of investigators have yielded a detailed picture of certain important aspects of this process; in particular, the control of ovalbumin synthesis by estrogenic hormones. As will be apparent from the specific cases to be discussed, the possibility of regulation exists at several levels: (1) control of the rate of transcription; (2) control of the rate of degradation of already synthesized messenger; (3) posttranslational regulation by modulation of the rate of translation on the ribosomes; (4) activation of an already synthesized enzyme through the intervention of another protein that is itself induced by hormones.

II. RESPONSE TO STEROID HORMONES IN EUKARYOTIC CELLS

As has previously been discussed, there are a variety of molecular mechanisms by which gene expression could be controlled in eukaryotic cells. One modulatory pattern of gene expression that has been the subject of much consideration is hormonal response. Hormones, which regulate a wide spectrum of cellular behaviors, can be divided into two major groups: (1) steroids, which work directly at the level of the gene, and (2) proteins, including polypeptides, which in many instances operate through the mediation of cyclic AMP. We will first consider the properties of steroid hormones.

Much evidence indicates that the major regulation of ovalbumin synthesis by steroid hormones in the chick oviduct proceeds by regulation of the rate of mRNA synthesis (O'Malley and Schrader, 1976). Cells have receptor proteins that bind steroids; binding results in the formation of activated complexes which pass into the nucleus and bind to specific acceptor sites. O'Malley and his colleagues (1977) showed that ovalbumin synthesis in the chick oviduct is due to new messenger synthesis. This was demonstrated through the use of a cell-free system that contained RNA extracted from hormonally treated or untreated cells. Such a system synthesizes ovalbumin only in the case of the messenger from the estrogen-treated animals. Schimke and his collaborators (1973) have studied the same system and have measured ovalbumin mRNA and ovalbumin-synthesizing polysomes. They have also concluded that the rate-limiting step in ovalbumin synthesis is the quantity of specific messenger; however, their investigations do not allow them to distinguish between synthesis and stability of messenger as the mechanism responsible for its elevation.

Studies following the movement of labeled estrogen into the cell have identified specific protein receptor molecules that bind estrogen. Competition experiments indicate that these binding sites in the cytoplasm for estrogens number in the thousands. Upon combining with the estrogen molecule, a hormone–receptor complex is formed and migrates into the nucleus.

The efforts of O'Malley and his collaborators (Woo et al., 1978) to define the precise mode of regulation of the ovalbumin gene through DNA cloning methods have produced tantalizing results. These authors have developed a recombinant plasmid that contains a cDNA insert coding for ovalbumin (referred to as pOV230). However, analysis of the native gene by restriction analysis showed that the structural gene is noncontiguous and that at least two "introns," or noncoding sequences, were believed to exist within the ovalbumin gene (Fig. 11.1). Subsequently, it was determined that the gene was in fact much more complex and in a natural state contained seven intervening sequences that did not code for protein in the final product. Therefore, the formation of a functional mRNA must involve transcription, RNA splicing and base modification, transport from the nucleus, and assembly into polysomes (Wickens et al., 1980).

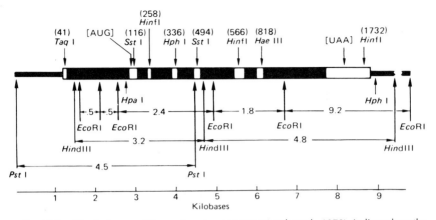

Fig. 11.1. Physical map of the ovalbumin gene (Dugaiczyk et al., 1979). Indicated are the key restriction cleavage sites, initiation and termination codons, and interspersed structural (thick open line) and intervening sequences (thick shaded line).

Although neither the significance of these sequences nor their relationship to the normal regulation of the gene is yet understood, it is clear at this point that the structural sequences are coordinately controlled in the response to estrogens.

In addition to pOV230, other plasmids that contain fragments of this ovalbumin gene have been developed. These have been used to examine the mechanism by which transcription is initiated. Such plasmids were sources of DNA for transformation into mouse cells (Lai *et al.*, 1980) and for injection into frog oocytes (Wickens *et al.*, 1980). It was observed that plasmid pBR322 containing a native ovalbumin gene with accompanying introns and flanking DNA sequences was capable of producing ovalbumin in mouse fibroblasts and in *Xenopus* oocytes. However, the much simpler gene pOV230 functioned equally well or better in the oocyte system. Because pOV230 contains neither introns, flanking DNA, cap site, polyadenylation site, nor TATAT sequence at the 5' terminus, none of these appear to be required for optimal transcription of the gene. Thus, many questions in the control of ovalbumin transcription remain unanswered.

In addition to control of mRNA synthesis, steroid hormones can also modulate gene expression by controlling the rate of translation of messenger (point 3 in Section I). Palmiter (1972) has demonstrated that treatment of chicks with estrogen results in a 3-fold increase in the number of ribosomes translating messenger and a 30 to 50% increase in the rate of peptide elongation on the ribosome. These observations have been made for the synthesis of egg white proteins in the oviduct, including ovalbumin, conalbumin, ovamucoid, and lysozyme. Thus, effecting the rate of transcription may be the most dramatic, but certainly not the only, mechanism by which steroids can modulate gene function.

Although the most economical fashion by which a cell could bring about the discontinuous production of gene products would appear to be through transcriptional controls, other possibilities exist. Although apparently wasteful, an enzyme could theoretically be synthesized and held in reserve until activated by intracellular signals. Such a mechanism has been described in the case of alkaline phosphatase. Alkaline phosphatases (EC 3.1.3.1) (see discussion on BUdR) are a heterogeneous group of enzymes of unknown physiological function (orthophosphoric monoester phosphohydrolases). At least three different human alkaline phosphatases coded by different genes have been identified: one that is characteristic of placenta (PAP), one that is related to the liver (LAP) (Badger and Sussman, 1976), and one that is related to the intestine (Mulivor *et al.*, 1978). The enzymes are present mostly in the nucleus and are easily assayed using chromogenic substrates (*p*-nitrophenyl phosphate). The enzyme has been intensively studied and varies widely from cell line to cell line (Cox and MacLeod, 1962).

Cell lines with low constitutive activity of alkaline phosphatase possess a single, heat-stable enzyme, whereas those cell lines with high activity possess two enzymes: a heat-stable type and a heat-labile type (Hertz, 1973). The heat-labile enzyme (found in HeLa S1 cells) appears to be similar but not identical to the PAP form (Elson and Cox, 1969). Those cell lines with low basal levels are 10- to 20-fold inducible with glucocorticosteroids, such as cortisol and dexamethasone. The addition of actinomycin D inhibits induction if added at the same time as the hormone but not if added 18 hours later (Griffin and Cox, 1966); this finding suggests that mRNA synthesis is required for induction. Furthermore, the induction process is blocked by puromycin, which would indicate an involvement with protein synthesis.

The studies on the properties of the enzyme also argue that a new protein is being synthesized in response to the hormone. Thermal stability and immunological specificity of the induced and basal enzymes are different, as is the elution pattern on DEAE-cellulose. Experiments with synchronized cells indicate that the cells must go through the S phase for induction (Griffin and Ber, 1969). Although these results argue circumstantially that steroid hormones regulate alkaline phosphatase by controlling the rate of enzyme synthesis, this is not the case. Cox and King (1975) have shown that the amount of alkaline phosphatase protein is not changed in hormonally induced cells. They have reached the following conclusions: Enzyme induction involves a structural locus and a modifier locus. Steroids interact with specific cell receptor molecules and then activate the modifier locus (only during S phase of the cell cycle). The interaction of the modifier (which itself appears to be a phosphatase) with the alkaline phosphatase molecule causes an alteration in zinc binding, and this effect lowers the energy requirements of the enzyme substrate transition state. In fact, Griffin and co-workers have shown that the induced enzyme binds one-half the number of phosphate groups compared to the number bound to the uninduced enzyme, resulting in altered catalytic activity (Bazzell *et al.*, 1976).

Despite this unequivocal demonstration of an indirect mechanism by which the steroid exerts its effect, there appears to be an example of an alternative inductive mechanism. Hamilton *et al.* (1979) have shown that in the induction of alkaline phosphatase by BUdR and dibutyryl cyclic AMP in two human carcinoma cell lines, *de novo* synthesis of alkaline phosphatase is occurring in both cases.

An alternate mechanism by which gene regulation could be controlled by hormones is through the control of mRNA breakdown. This was suggested to be the means by which the activity of the liver-specific enzyme tyrosine aminotransferase is modulated. Tyrosine aminotransferase is a tissue-specific enzyme that can be induced in liver or cultured hepatoma cells by adrenal steroids (Chapter 7). When presented with hormone, there is a dramatic increase in specific activity, reaching a maximum after about 10 hours. The induced activity is maintained by continuous exposure to the hormone; or, after removal, it will return to the basal level (Fig. 11.2). There is an increased rate of enzyme synthesis and new enzyme molecules are made, i.e., immunoprecipitation studies have shown that enzyme synthesis occurs rather than an activation of already existing molecules.

The Tomkins model (Tomkins *et al.*, 1969) proposed two genes involved in TAT synthesis: G^s, the structural gene, and G^r, a regulatory gene. According to this model, the inducing steroids antagonize a posttranscriptional repressor produced by the G^r gene. This posttranscriptional repressor has the dual function both of inhibiting TAT messenger translation and of promoting messenger degradation. The repressor is assumed to be a protein, and both the repressor and its messenger must be very labile relative to the other molecules. This model has been proposed as a general regulatory scheme for a wide variety of inducible systems. It satisfactorily explains a number of features of this mode of enzyme induction.

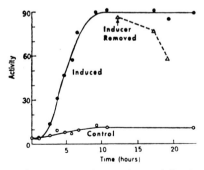

Fig. 11.2. Induction of tyrosine aminotransferase following exposure to glucocorticosteroid. (From Tomkins *et al.*, 1969.)

1. The steroid stimulates enzyme synthesis.

2. Enzyme-specific mRNA accumulates even when protein synthesis is inhibited.

3. A constant presence of the inducer is required to maintain the induced rate of enzyme synthesis.

4. The synthesis of RNA is required (i.e., induction will not occur if actinomycin D is presented simultaneously with the steroid).

5. If RNA synthesis is blocked after full induction, enzyme synthesis continues at the induced rate.

6. Actinomycin D, when presented to the cells following induction, "superinduces," causing an increase in intracellular TAT concentration. Superinduction is due to an increase in the rate of enzyme synthesis promoted by inhibiting RNA synthesis.

Although a model of regulation based on control of the rate of mRNA breakdown explains a number of features of the system, it must be viewed with skepticism, since it was based mainly on evidence obtained indirectly through the use of inhibitors. Such antibiotics are known to have multiple, complex actions, and more recent studies suggest that transcriptional regulation is responsible for variations in the level of TAT (Granner *et al.*, 1979; Kenney *et al.*, 1979). Using an *in vitro* translation assay it is possible to quantify mRNA production of TAT messenger: it was shown that following the addition of hormone there was a rapid increase in the level of TAT messenger, followed after a 1-hour lag period by an increase in the level of the enzyme. Furthermore, "reinduction" (i.e., induction brought about by subsequent addition of hormone) resulted in an increase both of TAT enzyme levels and TAT messenger. Kenny *et al.* (1979) reached similar conclusions based on the use of a *Xenopus* translation system in which guanidinium chloride-extracted liver mRNA was assessed by microinjection into oocytes. Thus, improvements in the technology of TAT messenger measurements would tend to support the belief (Kenney *et al.*, 1973) that superinduction results from stabilization of mRNA by actinomycin D. However, posttranscriptional control may be important in other systems, and its possible role should not be ignored (Sato and Ross, 1979).

III. EFFECTS OF BROMODEOXYURIDINE ON DIFFERENTIATED GENE FUNCTIONS

The study of differentiation can be expedited using specific chemical agents that affect development. One of the most widely used substances presently available is 5-bromodeoxyuridine which is incorporated into DNA in place of thymidine, causing an increase in buoyant density of the DNA in a CsCl gradient

(Luk and Bick, 1977). Resistance to BUdR has been an extremely useful genetic marker and has been widely employed in cell hybridization studies (see Chapter 2). Furthermore, BUdR substitution can lead to increased radiation sensitivity and can be mutagenic (see Chapter 12). However, our concern here will be its inhibition or induction of a variety of cellular functions associated with differentiation (Table 11.1).

The general features (Rutter *et al.*, 1973) of the BUdR effect on eukaryotic cells are:

1. The blockage of specific protein synthesis occurs at a concentration of BUdR that does not substantially affect total synthesis of DNA, RNA, or protein.
2. The effect is blocked by thymidine.
3. The treatment with BUdR must take place during S phase in order for the effect to be manifested.

TABLE 11.1   Some Examples of the Effects of Bromodeoxyuridine on Various Cells[a]

System	Parameter measured
Repression	
Pancreatic acinar cells	Exocrine enzymes
Pancreatic B cells	Insulin
Chondrocytes	Chondroitin sulfate
Myoblasts	Myotube formation, myosin
Pigmented retina cells	Melanin
Erythroblast precursors	Hemoglobin
Mammary gland	Casein, α-lactalbumin
Lymphocytes (primed)	Antibodies
Amnion cells	Hyaluronic acid
Liver cells (avian)	Estrogen induction of phosvitin
Hepatoma cells (HTC)	Glucocorticoid induction of tyrosine aminotransferase
Mouse lung	Hyaluronic acid
Mammary carcinoma hybrid cells	
Melanoma	Tumorigenicity, pigments
Induction	
Mouse lung/mammary carcinoma hybrid cells	Alkaline phosphatase
Mouse mammary carcinoma cells	Alkaline phosphatase
Pancreatic exocrine cells	Alkaline phosphatase
Neuroblastoma	Neurite formation: cell membrane glucoprotein
Lymphoid cells (Burkitt lymphoma clones and NC37 line)	Epstein–Barr virus

[a] From Rutter *et al.* (1973).

4. The recovery period varies depending on treatment.
5. The effect is not mutagenic; altered proteins are not produced.
6. The effect in general appears to be reversible (but exceptions exist).

Although the most obvious fashion by which nucleic acid analogs could affect gene expression would be through incorporation into DNA, a number of studies argue against this simple interpretation. One striking example is that neuroblastoma differentiation (Schubert and Jacob, 1970) has been shown to respond to the presence of BUdR (Table 11.2). The effect of BUdR is to induce neurite formation, and this effect is not blocked by a number of inhibitors of DNA synthesis. This would suggest that BUdR incorporation into DNA is not responsible unless a low-level incorporation into a minority fraction is involved.

The fact that the properties of the BUdR-induced neuroblastoma differentiation do not appear to follow the general rules suggested by Rutter *et al.* (1973) may indicate that in this situation the analog does not work through incorporation into DNA. An alternative mechanism involving modifications of the cell membrane through inhibition of glycosyltransferases has been suggested (Rogers *et al.*, 1975). Such enzymes are responsible for the biosynthesis of surface glycoproteins and glycolipids, and sugar donors such as UDP glucose could be interfered with by pyrimidine analogs. The observation that chick fibroblasts grown in BUdR become agglutinable by concanavalin A (see Chapter 2) would support the idea that alterations of sugar molecules under those conditions are responsible for changes in gene expression (Biquard, 1974). However, the evidence for such a scheme is circumstantial and the model is largely speculative.

In the permanent rat skeletal myoblast cell line L6 (see Chapter 8), BUdR prevents fusion and differentiation into myotubes (Rogers *et al.*, 1975). However, when deoxyuridine, deoxycytidine, or thymine are added simultaneously with BUdR, differentiation proceeds. Because these three substances do not prevent the incorporation of BUdR into DNA, BUdR may be acting at a different

TABLE 11.2   Inhibition of Neurite Formation[a]

Inhibitor[b]	Concentration	Differentiation (%)	Viable cells (%)
Puromycin	10 μg/ml	4	72
Cycloheximide	40 μg/ml	7	80
Actinomycin D	5 μg/ml	73	78
Vinblastine sulfate	$4 \times 10^{-6}\ M$	2	89
Colchicine	$2 \times 10^{-6}\ M$	81	87

[a] From Schubert and Jacob (1970).
[b] Cells were grown in $4 \times 10^6\ M$ BUdR for three generations in petri dishes and then plated on tissue culture dishes in the presence of the inhibitors. After 15 hours the percentage of differentiated cells and cell viability was determined.

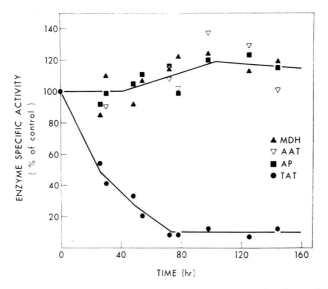

Fig. 11.3. Effect of BUdR on enzymatic activities in cultured cells (Stellwagen and Tomkins, 1971a). Cells in logarithmic growth were maintained in BUdR-containing medium. MPH, Malate dehydrogenase; AAT, alanine aminotransferase; AP, acid phosphatase; TAT, tyrosine aminotransferase.

site to suppress differentiated gene function. The authors hypothesize that the inhibition of glycosyltransferases by BUdR might affect the biosynthesis of fusion-specific glycoproteins, thereby affecting a variety of differentiated functions. However, in another study by Morrill *et al.* (1980) it was observed that a thymidine kinase-deficient mouse myoblast line was unable to incorporate BUdR into its DNA and failed to respond to the differentiation-suppressing effects of BUdR. Such findings suggest that BUdR must be converted to bromodeoxyuridine monophosphate in order to affect differentiation and that incorporation into DNA is the mechanism by which differential gene expression is suppressed.

Davidson and Kaufman (1977) have shown that the effect of BUdR on the inhibition on melanin synthesis can be reversed through the addition of deoxycytidine, a finding that is comparable to the results of Rogers *et al.* (1975). At the concentration range where the greatest effect occurs there is essentially no change in the amount of BUdR incorporated into DNA, indicating that the analog's effect is through other cellular systems.

In contrast to the previously cited studies, numerous investigations have endorsed the idea that BUdR acts through direct incorporation into DNA. For example, the response of the HTC cell line to BUdR has been the subject of several investigations (Stellwagen and Tomkins, 1971a,b). The results have shown (Fig. 11.3) that under conditions in which the activity of the liver-specific enzyme tyrosine aminotransferase is suppressed by 90%, there is no effect on several other enzymes. However, the effect of BUdR appears to be more general than simply an extinguishing of differentiated gene functions, because some "housekeeping" functions (i.e., glucose-6-phosphate dehydrogenase and lactic dehydrogenase) also appear to be suppressed. Furthermore, both the inducibility of TAT and its low-level basal activity are shut off by BUdR. There is a good inverse correlation between TAT activity and incorporation into DNA, suggesting that the mode of action of BUdR is through its incorporation in place of thymidine (Fig. 11.4). Experiments using synchronized cells have shown that the BUdR effect requires the presence of BUdR during the S phase of the cell cycle (Fig. 11.5). This is reminiscent of the observation that in the λ phage system transcription is inhibited in BUdR-substituted viruses.

Globin messenger production (Chapter 8) is also affected by BUdR. Friend leukemia cells treated with dimethyl sulfoxide are stimulated to make hemoglobin, and there is a concomitant increase in the production of mRNA for globin (Table 11.3). Bromodeoxyuridine treatment of such DMSO-stimulated cells results in a decrease of both globin and globin messenger production. Similar observations have been made in bacteria, in which a stimulatory effect is noted in the *lac* operon when DNA is BUdR substituted. Apparently, the incorporation into DNA results in a change in the physicochemical properties of the operator locus and a tighter binding of the repressor molecule. The rate of repressor dissociation is, in fact, 10 times slower in substituted than in unsubstituted DNA

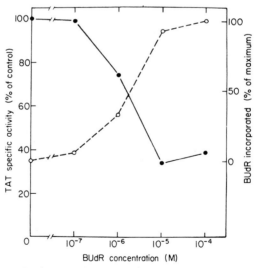

Fig. 11.4. Relationship between the incorporation of BUdR into DNA and TAT activity. ○, Incorporation of radioactive BUdR into DNA; ●, TAT specific activity (from Stellwagen and Tomkins, 1971a).

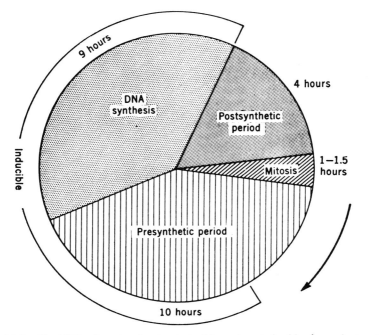

Fig. 11.5. The HTC cell cycle, showing inducible and noninducible phases for tyrosine aminotransferase (Tomkins *et al.*, 1969).

(Lin and Riggs, 1972). These findings suggest a mechanism by which the alteration of differentiated functions may occur. In support of the applicability of this hypothesis to eukaryotes is the observation that a mutant Friend leukemia cell line that lacks thymidine kinase and is unable to incorporate BUdR into DNA fails to show inhibition of hemoglobin synthesis (Ostertag *et al.*, 1973).

Chick embryonic limb bud cells can be cultivated for short periods *in vitro* and

TABLE 11.3   Effect of BUdR on the Amount of Globin mRNA in
Total Cell RNA of Friend Leukemia Cell Cultures[a]

Culture conditions	RNA hybridized to cDNA per total cell RNA (%)
Control	0.00025
+BUdR	0.0011
+(CH₃)₂SO	0.146
+BUdR + (CH₃)₂SO	0.039

[a] Relative amounts of globin mRNA in total cell RNA from Friend leukemia cells. Modified from Preisler *et al.* (1973).

will differentiate into cartilage in monolayer culture. The cartilage differentiation (Levitt and Dorfman, 1972) is repressed irreversibly by BUdR, as characterized by cellular morphology, metachromasia, and sulfate incorporation into acid mucopolysaccharide. The repression by BUdR could be reduced by addition of thymidine simultaneously but not subsequently further substantiating the irreversibility of the mechanism. These results suggest an interruption of a differentiation "program" requiring a particular time-ordered sequence of events.

Alkaline phosphatase (Bulmer *et al.*, 1976) can be either induced or repressed by BUdR, depending on the cell line. The effect is reversible and appears to be similar to that of glucocorticosteroids in that cell lines with high basal activity are repressed and low-level lines are induced. The enhanced phosphatase activity appears to be due to induction of a placental-type, heat-stable enzyme, whereas a liver, heat-sensitive alkaline phosphatase is repressed (Hamilton *et al.*, 1979). When either BUdR or glucocorticosteroids are presented to the cell, there is a substantial lag period of as much as 24 hours before an increase in the specific activity of alkaline phosphatase can be detected. This observation suggests that an intermediate event is required in order to accomplish the induction process. However, if the drugs are given in pulses as short as 4 hours, followed by growth in drug-free medium, there is an increase in alkaline phosphatase activity as much as 100 hours later. Thus, the inducers set in motion a chain of events that persist after the removal of the inducers. Experiments with the DNA synthesis inhibitor hydroxyurea indicate that DNA synthesis is required for induction (Morrow *et al.*, 1979).

There have been a number of models put forth to account for the BUdR effect via incorporation into DNA. For instance, Strom and Dorfman (1976) have proposed a mechanism of action involving preferential incorporation of BUdR into regulatory sequences, whose half-life would be thereby decreased because of structural instability. Davidson *et al.* (1975) have shown that 80 to 100% of mRNA molecules transcribed in sea urchin embryos are made from DNA adjacent to interspersed repetitive sequences. This suggests that differentiation may involve amplification of a special, moderately repetitive DNA set with a regulatory function next to unique sequences specifying structural genes. Strom and Dorfman's results indicate that BUdR is preferentially incorporated into moderately repetitive sequences of DNA in differentiating chick limb bud cultures; furthermore, cartilage-derived DNA contains a greater number of such BUdR-enriched sequences that does DNA of undifferentiated tissues. These results have been extended to neural retina (Strom *et al.*, 1978) with results that also suggest a DNA-linked mechanism.

If we assume that BUdR is being incorporated into regulatory sites on the DNA and thereby preferentially affecting the rate of mRNA synthesis, the results of Kallos *et al.* (1978) suggest this might occur by altering the binding of specific low-molecular-weight effector molecules to the DNA. These authors have shown that the estrogen·receptor from rat uterus has an enhanced affinity for BUdR-substituted DNA compared to unsubstituted DNA.

Another mechanism by which BUdR might exert a regulatory effect has been proposed by Fasy *et al.* (1980). These authors have studied the binding of a number of different proteins to halodeoxyuridine-substituted DNA and have found that all the halogen-substituted DNAs have an increased binding affinity for H1 histone. Although the exact mechanism of this enhanced affinity is not known, it was observed that the H1 histones could recognize the dose but not the type of halogen substitution. This is in contrast to the *lac* repressor, which also shows altered binding to halogen-substituted DNA, but with the following order of preference: IUdR-substituted > BUdR-substituted > ClUdR-substituted > unsubstituted (Lin and Riggs, 1976). These results offer a specific molecular mechanism by which these substances may alter gene expression; however, the discrepancy between prokaryotic and eukaryotic systems leaves a number of issues unresolved.

In conclusion, there is no generally accepted mechanism to explain the induction and suppression of differentiated gene functions by BUdR. Although evidence for a DNA-mediated model exists, some contradictory data support the hypothesis that interference with the synthesis and arrangement of membrane-associated components brings about a change in the expression of differentiated systems. It may be that more than one mechanism will have to be invoked to account for the BUdR effect.

IV. CYCLIC AMP RESPONSE IN CULTURED CELLS

Adenosine 3′,5′-monophosphate (cyclic AMP) is an intracellular "second messenger" that mediates the cellular stimulation of a variety of hormones, including proteins, polypeptides, and enkephalins (Fig. 11.6). Whereas in bacteria cAMP can exert its effect through gene activation, in mammalian systems it operates in an indirect fashion. The hormones trigger specific receptors located in the membrane to stimulate the activity of adenylate cyclase (EC 4.6.1.1), which converts ATP to cAMP, affecting a variety of cellular processes. Adenylate cyclase, however, is not directly activated by the binding of the hormone to the receptor. Rather, a third protein, the G protein, mediates this effect. When the hormone–receptor complex interacts with the G protein, GTP is exchanged for GDP. In turn the GTP–G protein complex activates adenylate cyclase. The principal mechanism by which cAMP regulates cellular activities is through cAMP-dependent protein kinases which mobilize already synthesized enzymes. This process is believed to occur through a binding of cyclic AMP to a regulatory subunit on the protein kinase. The binding results in a dissociation of the regulatory and catalytic subunits from the protein kinase and activation of the catalytic

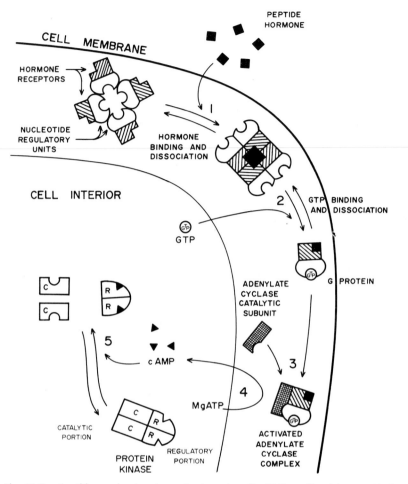

Fig. 11.6. Possible mechanism for activation of cyclic AMP mediated by peptide hormones. Peptide hormones bind with receptor–regulatory tetramer, changing its conformation (1). Complex can now bind GTP (2), resulting in dissociation of tetramers. Adenylate cyclase catalytic subunit is activated by its binding (3) to the GTP–receptor–regulator–hormone complex. cAMP is formed (4) and can bind to a regulatory subunit of the protein kinase, thereby activating the enzyme (5). Modified from Rodbell et al. (1981) and Baxter and Funder (1979).

portion. Two protein kinases exist, type I and type II, which have a common catalytic subunit (C) but which possess different regulatory subunits (RI and RII). Thus, the two major protein kinase holoenzymes are $(RI)_2C_2$ and $(RII)_2C_2$. Cell culture approaches have been useful in resolving the specific contribution of different macromolecules to this system (Gottesman, 1980).

When cells are presented with nonpolar analogs of cAMP, such as dibutyryl cyclic AMP (Bt$_2$cAMP), there is a dramatic series of responses, including retardation of cell growth, morphological alterations, decreased agglutinability by lectins, and altered amino acid and sugar transport (Hsie and Puck, 1971; Singh et al., 1981) (Fig. 11.7). Furthermore, many cell lines are highly sensitive to killing by cAMP. In order to understand the mechanism of these effects, a number of mutants resistant to the toxic effects of cAMP have been isolated. Some of these mutants have altered cAMP-dependent protein kinase activity.

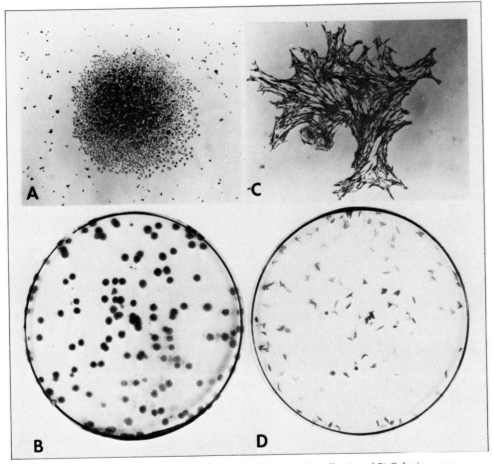

Fig. 11.7. Effect of Bt$_2$cAMP on the morphology of CHO cells. (A and B) Colonies grown from single CHO cells on standard medium for 7 days. ×42.5 and ×1.7, respectively. (C and D) Colonies grown in the presence of 1 mM Bt$_2$cAMP. ×42.5 and ×1.7, respectively. (Hsie, 1981.)

This finding has been confirmed in several instances (Simantov and Sachs, 1975; Masui *et al.*, 1978; Gutman *et al.*, 1978; Singh *et al.*, 1981). In the latter case, the mutants have no detectable type I protein kinase, indicating that although type I protein kinase is required for regulation of cellular activity, it is not essential for growth.

Little information is available concerning the properties of the proteins phosphorylated by the cAMP-dependent protein kinases. LeCam *et al.* (1981) have studied a number of cAMP-resistant CHO mutants in an effort to analyze this question. It was found that a 52,000 MW protein was consistently phosphorylated in wild-type cells following activation of cAMP-dependent protein kinases. This phosphorylation did not occur in the class of mutants lacking type II kinases. Although the role of this protein in cellular metabolism is not known at present, its phosphorylation appears to be crucial for sensitization to cAMP. However, other cAMP-resistant mutants are not altered in their ability to phosphorylate this protein. Thus, the 52,000 MW protein is necessary but not sufficient to account for the effect of cAMP on CHO cells.

The genetics of the cAMP effect can also be approached through the isolation of cholera toxin-resistant mutants (Evain *et al.*, 1979). This agent exerts its toxicity by increasing adenylate cyclase activity. Resistant mutants are refractory to killing by cholera toxin and possess a protein kinase with an altered catalytic subunit. This mutation lowers the affinity of the catalytic subunit for the ATP substrate, thus decreasing the response of the cells to cholera toxin and related agents. Through high-resolution, two-dimensional electrophoresis and affinity chromatography of the cAMP binding subunits of cAMP-dependent protein kinases, the molecular alterations of some of these mutants have been localized. Steinberg *et al.* (1977) have characterized a cAMP-resistant mutation of S49 lymphoma cells that is due to a single amino acid substitution in the R subunit. Yet another resistant class may be the result of a regulatory mutation. Steinberg *et al.* (1978) have described mutations of S49 lymphoma cells that behave as dominant traits in somatic cell hybrids. This transdominant effect could be explained by a mutation in a regulatory gene producing a diffusible repressor (Chapter 7).

In addition to mutants of the protein kinase structural genes, variants involving alterations of the enzyme adenylate cyclase have been described (Bourne *et al.*, 1975). These are capable of growth in isoproterenol, a stimulator of adenylate cyclase, and in the presence of a second drug that inhibits phosphodiesterase. A fourth class is represented by variants resistant to terbutaline, a β-adrenergic antagonist. The interaction between the hormone binding to the β-adrenergic receptors and the activation of the adenylate cyclase has been lost, and cells are now resistant to the toxic effects of this drug (Haga *et al.*, 1977). This is due to the failure of the hormone to stimulate cyclase activity, even though the receptor and cyclase are both present.

V. CONCLUSIONS

Gene expression in eukaryotic cells can be modulated through the addition of a variety of low-molecular-weight substances. One of the most widely studied of these is BUdR, the mode of action of which is still obscure. Because of the conflicting lines of evidence, it appears likely that this substance acts through several different mechanisms. These may include disruption of normal membrane glycoprotein formation and direct intervention through incorporation into DNA. Several mechanisms have been proposed by which the latter might occur, including interference with repressor binding or dissociation, or a preferential degradative effect on DNA segments (regulatory elements) involved in gene regulation.

A great deal is known concerning the molecular mode of action of steroid hormones, which includes complexing with specific cytoplasmic receptors and activating gene expression through binding of the receptor–hormone complex acceptor sites on the DNA. The precise relationship of the intervening sequences in structural genes to regulation is not yet clarified, but the use of native and artificial DNA probes is being employed in order to analyze hormone action. Although steroid hormones may increase gene expression through induction of new messenger, they can also operate by several other mechanisms, including activation of already synthesized protein and stabilization of messenger RNA.

Finally, somatic cell genetics is being used to clarify the mechanism by which cyclic AMP produces its effect. A number of mechanisms of resistance to the cytolytic effects of this substance have been clarified, including modifications in both adenylate cyclase and protein kinase. This approach should prove to be especially useful in enlarging our understanding of the intermediate reactions in the cyclic AMP pathway.

Mutation Induction in Cultured Cells

I. INTRODUCTION

Previous chapters in this volume have dealt with genetic variability in eukaryotic cells and its relationship to biological processes. Up to this time we have focused mainly on spontaneous genetic changes; the involvement of chemical and physical agents in mutagenesis has not been considered in detail. The importance of induced mutational change to health-related questions is evident; for example, it has been argued that a substantial percentage of malignancies may result from chemical and physical mutagens existing in the environment (Cairns, 1978). Similarly, atherosclerotic lesions may result from events related to cellular transformation, which could also arise from environmentally based mutagenesis (Chapter 13). In addition, pharmacological agents may introduce potential mutagenic hazards into the environment.

The process of drug development requires extensive testing for hazardous side effects, particularly for carcinogenic or mutagenic potential. Thus, testing procedures that are rapid and efficient are of great interest. Because every conceivable assay system has drawbacks, new techniques are constantly being evaluated. Cell culture is in some aspects closer to a relevant system for the measurement of mutations in man than bacteria, *Drosophila,* or yeast and is much less expensive than testing mammals for single-locus mutations. Moreover, experimental mutagenesis has proved to be virtually impossible to study in humans for both ethical and technical reasons. Despite 35 years of intensive analysis, even in the case of the atom bomb survivors of Hiroshima and Nagasaki, there is no unequivocal evidence for mutation induction in man as a result of exposure to ionizing radiation (Vogel and Motulsky, 1980). The children of exposed individuals have been monitored for an alteration in the sex ratio, which is a sensitive measure of either dominant or recessive sex-linked lethal mutations. The results to date have been negative; there is no statistically significant difference in male and female ratios among the progeny of the bomb survivors. Although a substantial increase in leukemia incidence among irraidated survivors argues for the carcinogenic effect and indirectly for the mutagenic potential, evidence for mutation induction in man is nonexistent.

A great body of information concerning mutagenesis in higher eukaryotic organisms has resulted from studies on corn, *Drosophila,* and mice. However, such investigations are limited by the lack of effective screening systems and often require the examination of thousands of plants or animals. Therefore, mammalian cell culture using suitable selective markers (Chapters 1–3), combined with very large numbers of individual cells, should prove to be significant in the study of mutagenesis.

Early investigations employing mutagenic agents in cultured cells gave negative results, resulting in some confusion. In recent years, these early problems have been resolved through the use of better selective systems and through a better understanding of the process of fixation and expression of mutational damage in eukaryotic cells.

In bacteria, the importance of DNA repair to the mutational process has been reasonably well established. It has been demonstrated that two repair processes—''error-free'' and ''error-prone''—enzymatically repair damaged DNA, with mutations being generated in the latter process (Witkin, 1976). In mammals, there also appears to be at least two repair processes for mutational damage. One of these is excision repair, which involves the enzymatic removal of pyrimidine dimers and their subsequent replacement. This repair system is lacking in xeroderma pigmentosum (Chapter 13) and is apparently error-free. A second repair process is an error-free, postreplication process. Finally, there may exist an error-prone, postreplication process in mammalian cells (Chang *et al.,* 1977).

II. PROBLEMS INVOLVED IN MUTAGENESIS STUDIES

A. Predisposition of Cells to Treatment

Mammalian cells are heterogeneous with respect to their genetic and metabolic state. These sources of variation can best be controlled by careful attention to variables that could affect the response of cells to mutagens. These include storing cells frozen in ampules, cloning of populations, standardizing of culture conditions, and careful monitoring of sera, which could be a major source of variation. Frequent cloning of cells is important because mutants can constantly accumulate in cell populations (Chapter 1), a process that can significantly alter the spontaneous mutation frequency. These factors may be responsible for the substantial variation that has been seen in various experiments (Thacker *et al.*, 1976).

B. Determination of Expression Time

When cells are exposed to mutagenic agents, newly induced mutations may not immediately display their altered phenotype but may require a period of time for maximum expression; this phenomenon has often been called phenotypic delay. The expression of new induced phenotypes is probably dependent mainly on a period required for fixation of the mutation in DNA, plus the disappearance of old gene products from the cell, or the need to accumulate altered gene products. Distinguishing these various factors is a complex problem, because during the expression period spontaneous new mutations can occur and add to the number induced by the treatment. In addition, the mutant growth rate may be different from that of wild-type cells, both for established and newly induced mutations (Fig. 12.1). In the case of thioguanine resistance induced by ethyl methane sulfonate (EMS), O'Neill and Hsie (1979) have demonstrated a lag period of 7 to 9 days for maximal expression. This lag period occurs with successive subculturing or when cells are maintained in a viable, nondividing state in serum-free medium; thus, continued cell division is not essential for the expression process. The expression period can vary tremendously in different systems. Using mouse lymphoma cells and thioguanine resistance, Sato and Heida (1980) have observed a period of 14 days to be required for maximum expression of uv-induced mutations. Thus, it is not surprising that some early

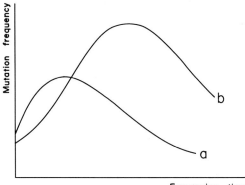

Fig. 12.1. Frequency of radiation-induced mutants as a function of expression time in V79 Chinese hamster cells at two doses of uv or X rays. a, Low dose; b, higher dose (Simons, 1974).

studies gave negative results, because a lag of this magnitude would hardly have been predicted.

C. Recovery of Mutants

Are all mutations caused by the agent recoverable? Many times, no—because of the influence of metabolic cooperation (Chapter 2). Because metabolic cooperation is especially pronounced for purine analog resistance in normal diploid cells, screening for mutants in such systems is a laborious task and requires large numbers of petri dishes in which the cells are plated at low enough densities to avoid cross-cooperation (O'Neill *et al.*, 1977). Some systems, such as ouabain selection, do not appear to be influenced by metabolic cooperation. In this case, a decline of mutant yield at high cell densities is observed, but this may be related to starvation from overcrowding (Arlett *et al.*, 1975).

D. Heterogeneity of Induced Variants

Much of the ambiguity and variability encountered in earlier studies is no doubt due to the sources of confusion that have plagued studies on spontaneous mutagenesis. The fact that many mutant (drug-resistant) phenotypes are heterogeneous (Chapters 1–3) requires a careful characterization of a substantial sample of putative induced mutants. This precaution has been adopted as standard procedure in many laboratories and may be largely responsible for a notable improvement in the quality of data collected in mutagenesis studies.

E. Activation of Mutagens

Some molecules must be converted to active forms in order to induce mutational damage. In cell culture systems, methods have been developed in which tumor cells are inoculated into an appropriate animal and, after treatment with the presumptive mutagen, are reintroduced into culture in the presence of the selective agent. Such a procedure provides an opportunity for the liver to enzymatically transform molecules into forms that are mutagenically active (Gabridge and Legator, 1969). Other variations include the microsomal activation system (Malling, 1971) and the cell-mediated assay (Huberman and Sachs, 1974).

III. INDUCTION OF MUTATIONS BY X RAYS AND NONIONIZING RADIATION

Hamster cells have been frequently used for the study of uv-induced mutations at the HPRT gene. Arlett and Harcourt (1972) have shown that the optimum expression time for new mutations is a function of dose, and the greater the dose, the longer the lag period for mutant expression, as if the mutated cells were receiving more damage (Fig. 12.1). Bridges and Huckle (1970) performed similar experiments, which showed a linear relationship between dose and number of mutants, as did the experiments of Hsie *et al.* (1975). The mode of action of uv is not clear. Although the predominant biochemical effect of uv is the production of thymine dimers, it is likely that its role in the production of mutations involves other mechanisms as well. Zelle *et al.* (1980) have shown that uv irradiation in the greater-than-290-nm range was more effective on a per dimer basis in inducing mutations than 254-nm radiation. This excess at the longer wavelengths suggests that other biological damage is also involved in the production of mutations.

The fact that mammalian cells possess a distinct phase of their life cycle during which DNA is replicated suggest that they might be expecially sensitive to mutagenesis during this period. This belief is based on models of mutagenesis developed in bacteria and bacterial viruses and assumes that errors in the process of replication can result in the insertion of fraudulent bases. If the period during which an individual gene is replicated is fixed and heritable, then during a small fraction of the S phase it should be possible to obtain a large yield of mutations at the replication period of that particular gene. Thus, by irradiating synchronized cells at different intervals or by pulsing cells with BUdR followed by treatment with visible light, it should be possible to determine at what point a particular gene is replicated. By using these approaches it has been shown that a unique period of high sensitivity to mutagenesis exists for the *HPRT* gene (Riddle and Hsie, 1978) and the *Ala-32* gene (Suzuki and Okada, 1975). However the Na^+/K^+-*ATPase* gene (ouabain resistance) shows two periods of mutagenesis susceptibility, in early and late S phase (Tsutsui *et al.*, 1981). This finding suggests the existence of more than one locus or of more than one mechanism of mutagenesis for this genetic marker.

As previously mentioned (Chapter 1), not only uv but also near-uv and visible wavelengths can be mutagenic to cultured cells. Hsie *et al.* (1977) have tested fluorescent white and black light, sunlamp light, black-blue light, and actual sunlight exposure. All were substantially mutagenic by the criteria of the thioguanine/CHO screening system. Similar results were obtained by Jacboson *et al.* (1978) in the BUdR/mouse lymphoma cell assay. Exposure to fluorescent light during the growth period also results in the appearance of new mutations (Jostes *et al.*, 1977), possibly as a result of the production of mutagenic photoproducts from tryptophan and riboflavin in the growth medium (Wang, 1976).

Theoretically, it is to be expected that ionizing radiation, which is much more energetic than uv, would produce a variety of genetic alterations. X rays and gamma rays are photons of energy that are energetic enough to eject electrons from their outer shells and thereby to cause the production of positive ions.

Positive and negative ions and free radicals would be expected to bring about a variety of biological alterations within the cell. This might include not only direct damage to the DNA but also secondary damage caused by reactive intermediates. Thus, much more heterogeneity in mutagenic damage due to ionizing radiation would be anticipated. Chu (1971), using X rays on Chinese hamster cells, carried out experiments involving screening for azaguanine resistance; he established the conditions for demonstration of mutagenesis. The reversibility of many of the azaguanine-resistant mutants was tested by plating the cells in HAT medium after treatment with a variety of chemical mutagens: some were nonrevertable, some appeared to have base-pair substitutions, and some appeared to be frameshift mutants. Arlett and Potter (1971) have obtained similar results in which they found that synchronized cells in G_2 appear to be most sensitive and that splitting the dose lowers efficiency of mutagenesis.

In the case of human fibroblasts, the spontaneous mutation rate was 10^{-6} (Chapter 1; Albertini and DeMars, 1973). In X ray-induced mutants, a greater mutagenic effect at higher doses was noted and resulted in nonlinear kinetics. Metabolic cooperation is so great that relatively small numbers of cells (10^4 cells/dish) must be used; this procedure requires large numbers of petri dishes. The diploid fibroblast system is closer to *in vivo* systems, but cells are difficult to grow and technical problems such as low cloning efficiency make the system less useful. However, despite the many differences between permanent hamster

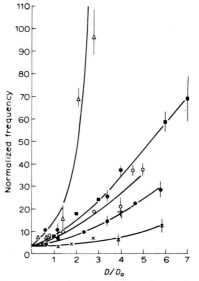

Fig. 12.2. Mutation frequency as a function of dose of X ray for five cell lines varying in their sensitivity to X rays. The sensitivity of the cell lines is directly related to their mutability ($\triangle$ > ■ > ○ > ● > ×) (Fox, 1974).

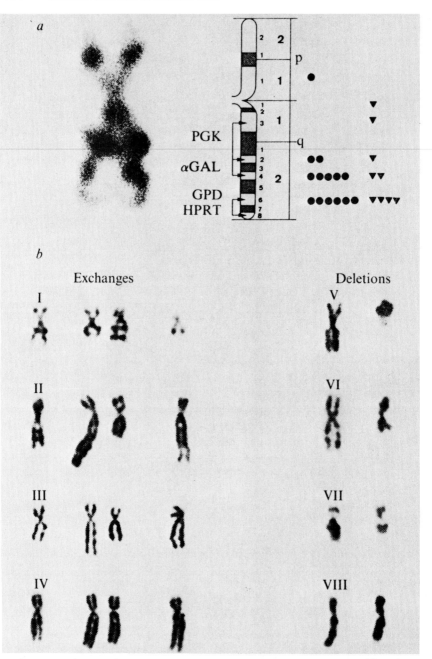

Fig. 12.3. Aberrant chromosome arrangements in thioguanine-resistant fibroblast clones induced by X rays. ●, Exchange points; ▼, deletion points; exchange and deletion points are

cell lines and human diploid fibroblasts, Thacker and Cox (1975) have shown that the mutation frequency of the two cell types is essentially identical when sensitivity to the inactivating effects of ionizing radiation is taken into account. Fox (1974) has examined the response of P388 mouse lymphoma cells to X ray-induced mutagenesis using excess thymidine as a selecting agent. In five different cell lines, it was found that there was a positive correlation between sensitivity of the cell line to X-ray killing and mutability (Fig. 12.2).

Studies by Cox and Masson (1978) indicate that at least 40% of HPRT mutants induced by ionizing radiation in cultured human fibroblasts are the result of chromosomal rearrangement (Fig. 12.3). This observation indicates that the rearrangement of blocks of genes may result in the suppression of particular loci (Chapter 1). These data are in agreement with those of Friedrich and Coffino (1977), who observed that X rays were effective in producing dibutyryl cyclic AMP-resistant mutants with reduced, but not altered, activity. A lowered activity would be in agreement with a model of mutagenesis resulting from gene rearrangement (Chapter 1), whereas an altered activity (i.e., K_m changes) would require a base-pair substitution. Furthermore, the fact that X rays (Friedrich and Coffino, 1977) and γ rays (Arlett *et al.*, 1975) were ineffective in inducing ouabain resistance (which is due to an alteration in the enzyme affinity, i.e., K_m, for the drug) is also in agreement with deletions and chromosomal rearrangements as the major mechanism of ionizing radiation-induced mutagenesis.

IV. CHEMICAL MUTAGENESIS

The first chemical mutagenesis studies in mammalian cells were carried out more than a decade ago (Kao and Puck, 1969; Chu and Malling, 1968). Subsequently, numerous experiments demonstrated that carcinogenic compounds (such as ICR-191 and ethyl methane sulfonate) are also highly mutagenic. Interestingly, caffeine, an effective producer of chromatid exchanges, does not cause mutations to nutritional auxotrophy. Since that time, numerous studies have established the mutagenic potential of many chemical agents.

O'Neill *et al.* (1977) have reported experiments on mutation induction to thioguanine resistance using ethyl methane sulfonate in hamster cells. The yield of mutations as a function of dose was linear (see Figs. 12.4 and 12.5). Their results could be fitted to the equation

$$f(x) = 10^{-6}(5.58 + 310.77x) \tag{12.1}$$

indicated on the diagrammatic X chromosome (A). Also indicated are the approximate locations of four X-linked genes. (B) Eight different mutants involving either exchanges (I–IV) or deletions (V–VIII). In I to IV the normal Xs are to the left, the normal autosomes to the right, and the exchanges are in the middle. In V to VIII, the normal Xs are to the left and the deletions are to the right. (From Cox and Masson, 1978.)

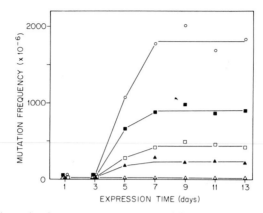

Fig. 12.4. Relationship between expression time of thioguanine-resistant phenotype and dose of mutagen. Cells were treated with 0 (△), 50 (▲), 100 (□), 200 (■), and 400 (○) mg/ml of EMS (O'Neill and Hsie, 1977).

Sato *et al.* (1972) studied chemical mutagenesis using azaguanine resistance as a marker in human lymphoblast cells. These cells are available from individuals with lymphoproliferative disorders as well as from healthy individuals, and they exhibit some of the properties of transformed cells. They possess an infinite life span, have a near-diploid karyotype, grow in suspension, and possess the Epstein–Barr virus. Ethyl methane sulfonate and methylnitronitrosoguanidine

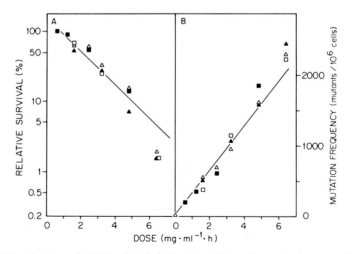

Fig. 12.5. Dosimetry for EMS cellular lethality (A) and induced mutation frequency (B). Cells were treated with various concentrations of EMS for 2 (△), 4 (▲), 8 (□), or 12 hours (■) (O'Neill and Hsie, 1977).

(MNNG) were both to be mutagenic in this system and to give nonlinear dose responses. Those mutants that were isolated were all deficient in HPRT, and the chromosome modes were close to 46.

The asparagine requirement of a number of mammalian cell lines has been employed in some chemical mutagenesis studies. The L5178Y mouse leukemia was used in a host-mediated system in order to examine the effects of an *in vivo* environment (Summers and Handschumacher, 1973). Because such cells do not grow as monolayers it is necessary to use soft agar for detecting clones. Three mutagens were employed: EMS, di-(2-chloroethyl)sulfide, and dimethylnitrosamine. They were mutagenic *in vivo,* but nitrosamine was not mutagenic *in vitro,* a result that is in accord with the belief that nitrosamine must be metabolized by the liver to an active form in order to be mutagenic. Using the asparagine-requiring Jensen sarcoma, Morrow *et al.* (1976) tested EMS quinacrine (a suspected antimutagen) and MNNG. Both EMS and MNNG are mutagenic, whereas quinacrine neither augmented nor decreased the mutation rate. An effective antimutagen would have a great potential value as an adjunct to chemotherapy because antimutagens would decrease the frequency of drug-resistant variants.

The relationship of carcinogenicity to *in vitro* mutagenesis has been a subject of substantial interest. Huberman and Sachs (1976) have tested a spectrum of compounds for mutagenicity. This work indicates that for the mutagenic effect as measured by mutation to ouabain resistance, azaguanine resistance, and high-temperature resistance, the potency of carcinogenicity for a given compound was related to the degree of mutagenicity. Further studies by Huberman *et al.* (1976) indicate that when a variety of metabolites of benzo[*a*]pyrene were tested, the most effective mutagenic agent is the non-K-region epoxide, diol epoxide, and that this substance may be responsible for the carcinogenicity of the parent compound. Positive results have also been obtained by Peterson *et al.* (1979) and Newbold and Brookes (1976) using V79 cells and azaguanine resistance in conjunction with highly carcinogenic polycyclic hydrocarbons. Those hydrocarbons must be activated in order to be mutagenic.

Orkin and Littlefield (1971) studied the mutagenic agent methylnitronitrosoguanidine, which acts at the replication fork of DNA in bacteria and causes mutations at the point of gene replication. The cells were synchronized and screened for mutations by using a double thymidine block. No correlation was found between the S phase and mutation induction (using BUdR and thioguanine resistance as genetic markers). These results also suggest that this chemical mutagen acts in an indirect fashion. Similar experiments were performed by Aebersold and Burki (1976) with BUdR mutagenesis in synchronized Chinese hamster cells. In this case there is a very sharp peak for mutation induction, indicating that BUdR interacts with replicating DNA to bring about induction of mutations.

Kao and Puck (1968, 1969, 1971), using a back-selection technique, have performed extensive screening for reversion of auxotrophic mutations. A variety of agents were tested and gave positive results (Table 12.1).

One of the most extensively investigated mammalian cell assays for chemical mutagenesis is the thymidine kinase system (Chapter 2). Clive and his collaborators have argued that mouse lymphoma cells possess two functional alleles (TK^+/TK^+) in the wild-type condition (Clive and Voytek, 1977). By selecting first with thymidine analogs, variants with no enzyme are obtained; these variants are assumed to be TK^-/TK^-. Revertants obtained in THMG medium are thought to be heterozygotes (TK^+/TK^-), and these are used to screen substances for mutagenicity. Thus, selection of such presumptive heterozygotes in trifluorothymidine requires only one mutational event in order to obtain enzyme-negative clones.

Such investigations, using a wide variety of carcinogenic substances (Table 12.2), show an excellent quantitative correlation between mutagenicity and carcinogenicity (Fig. 12.6). Clive and Moore-Brown (1979) have further argued that the mouse lymphoma system is a more appropriate method for predicting carcinogenicity than the Ames test (Chapter 10). They have observed that two types of resistant colonies are obtained when presumptive heterozygous cells (TK^+/TK^-) are plated in the selecting agent (trifluorothymidine). One class are small colonies, which may be simultaneously affected at the galactokinase locus that is linked to TK. These variants are possibly the result of position effects

TABLE 12.1   Calculations of Various Mutation Frequency Parameters Found for Forward Mutations in All Loci Studied by Several Chemical Mutagens[a]

Agent[b]	Summed overall mutation frequency for all loci tested ($\times 10^{-5}$)	Summed mutation frequency for all loci tested, dose equivalent to the D_0 value ($\times 10^{-5}$)	Summed mutation frequency for all loci tested adjusted to the 20% survival point ($\times 10^{-5}$)
None	0	—	—
EMS	5.0	2.8	9.8
MNNG	3.5	3.1	4.7
ICR-191	1.0	0.17	1.0
Hydroxylamine	0	0	0
Caffeine	0	0	0
uv	2.0	0.77	2.0
X ray	0.4	0.13	0.24

[a] From Kao and Puck (1971).
[b] Abbreviations for mutagenic agents as given in Table 12.2.

TABLE 12.2 Chemicals Tested in the L5178Y/TK$^{+/-}$ System[a]

Number	Chemical	Abbreviation
1	2-Aminopurine	2-AP
2	2-Acetylaminofluorene	2-AAF
3	4-Acetylaminofluorene	4-AAF
4	Benzo[a]pyrene	B[a]P
5	Benzo[e]pyrene	B[e]P
6	ϵ-Caprolactone	ϵ-Cap
7	Cyclophosphamide	CP
8	*p,p'*-DDE	DDE
9	Diethylnitrosamine	DEN
10	Diethylstilbestrol	DES
11	Dimethylnitrosamine	DMN
12	Diphenylnitrosamine	DPN
13	Ethylene dibromide	EDB
14	Ethyl methane sulfonate	EMS
15	Furylfuramide	AF-2
16	Hycanthone methane sulfonate	Hyc
17	Lucanthone hydrochloride	Luc
18–29	12-Hycanthone and lucanthone analogs	1A-3 to 1A-6, SW-1 to SW-8
30	Methotrexate	Mtx
31	Methyl iodide	MeI
32	Methyl methane sulfonate	MMS
33	*N*-Methyl-*N*-nitro-*N*-nitrosoguanidine	MNNG
34	Mitomycin C	MMC
35	Myleran	Myl
36	Natulan	Nat
37	Proflavin sulfate	Proflavin
38	β-Propiolactone	β-Prop; β-P
39	Saccharin, sodium	Sacch
40	Sodium azide	NaN$_3$
41	Succinic anhydride	Succ Anh
42	Uracil mustard	UrMus
43	Whole-smoke condensate	WSC

[a] From Clive and Moore-Brown (1979).

(Chapter 1) involving deletions of one or two bands because chromosomal re-arrangements involving chromosome 11 are observed in such clones (Hozier *et al.*, 1981). These observations would tend to confirm the suggestion (Morrow, 1977) that many mammalian cell variants result from gene inactivation. The second class are large colonies, and these appear to be the result of point mutations. Because both types of events (chromosomal and point mutational) are detectable, this system may be more reliable for the screening of potential carcinogens than methods that will detect only point mutations.

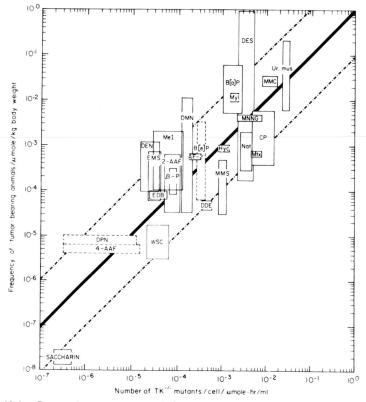

Fig. 12.6. Oncogenic potency in rats and mice versus mutagenic potency in the L5178Y system. Abbreviations are given in Table 12.2.

V. CONCLUSIONS

Notwithstanding that many problems in cell culture mutagenesis remain unsolved, including the mechanism and extent of repair, the biochemical nature of the lag period, and the failure of certain mutagens to follow their predicted mode of action, it is clear that great strides have been made in this area. Although a number of different systems have been described, none is ideal because the perfect mammalian cell mutagenesis assay would be one in which every single change at a particular locus could be detected. A cell capable of manufacturing a completely amino acid-sequenced gene product, such as human hemoglobin, would allow for a complete sampling of all mutations and a thorough description of all the genetic changes brought about by a particular agent. Although this would require an extensive chemical analysis, recent developments in the technology of DNA sequencing suggest that such an approach may be feasible. Thus, the examination of thousands of individual clones from a mutagen-treated population would allow the evaluation of mutagenicity at the primary level of the DNA. Although such an approach could at present be used to investigate mutagenesis mechanisms, the amount of effort required would be far beyond the resources of a screening program in which dozens or hundreds of potential mutagens are under consideration.

Perhaps the most challenging practical question with which animal cell mutagenesis studies will have to deal is the effect of very low level, chronic exposure to mutagenic agents. In human populations, extremely important legislative decisions regarding environmental pollutants will have to be made in the coming years. At present, most of our judgments regarding the potential hazards of those substances are based on extrapolation of data obtained from exposure of test systems to acute doses. Without knowledge of the effectiveness of repair systems at low exposure and of the possibility of a threshold for the effect of environmental mutagens, it will be impossible to make informed decisions concerning the levels of hazardous substances that are acceptable. If cell culture mutagenesis studies can contribute to a resolution of this problem, the investment of time and resources will be rewarded many times over.

Use of Cell Culture in the Analysis of Human Heredity

I. INTRODUCTION

Since the 1950s, when tissue culture procedures were improved, human genetics has taken advantage of the techniques of somatic cell genetics. Many significant findings in the genetics of man could have been made only with tremendous effort or would not have been uncovered at all without the aid of cell culture approaches. For example, our knowledge of the Lyon hypothesis owes a great deal to research on the properties of clones of fibroblasts grown from single cells. Similarly, much of our data on chromosomal mapping comes from somatic cell hybrids between human and animal cells.

Since many human genetic disorders are extremely rare, cell culture provides an alternative to laborious population screening in search of individuals or families that segregate for one or more genetic defects. Because questions concerning gene action and patterns of regulation are frequently studied through the use of complementation, it may be necessary to observe an extremely unlikely pedigree in order to determine the identity of two independent mutations. However, through the use of cell hybridization it is possible to combine two rare variants in the same cytoplasm. Because of the interaction through metabolic cooperation (Chapter 2) it is sometimes only necessary to cocultivate the cells. Thus, it has been possible to perform complementation studies using fibroblasts from patients with very rare metabolic disorders. For example, when fibroblasts from individuals with Hurler's syndrone are cocultivated with cells from a patient with Hunter's syndrome, the two mucopolysaccharidoses will mutually complement one another; this finding proves their dissimilar nature (Neufeld, 1974). The complementation is evidenced by a failure of cell mixtures to accumulate radio-

188

labeled mucopolysaccharides, a characteristic of the wild type cells and quite the converse of both mutant types, which pile up large amounts of these substances.

Despite considerable promise, cell culture has not been as successful as originally hoped in solving many of the problems in human genetics. This is the result of several technical limitations: (1) Human fibroblasts have a finite life span in culture (Chapter 14) and thus cannot be grown in unlimited quantities required for enzyme purification and analysis or for isolation of genes or regulatory macromolecules. (2) Culture medium is complex and, in most cases, poorly defined. No reliable substitute for animal serum has been obtained, and thus isolation and characterization of certain nutritional requirements is extremely difficult. (3) Fibroblasts cannot be grown in suspension; therefore their utility is limited for procedures requiring large volumes of material. (4) Cells derived from specialized adult tissue cannot be cultivated in a differentiated state *in vitro.* Thus, inborn errors of metabolism that are confined to specialized tissues (i.e., phenylketonuria; see Chapter 7) cannot be studied using this approach. Nonetheless, cultured cells have proved to be extremely useful in human genetics, as will be seen from a consideration of some specific examples.

II. HYPERCHOLESTEROLEMIA

Familial hypercholesterolemia, a genetic disorder in a specific cell receptor molecule, has been extensively studied *in vivo* and *in vitro,* and a coherent story for the genetics of this condition is emerging (Goldstein and Brown, 1979). This is a genetic disorder involving the processing of low-density lipoproteins (LDL): soluble lipoprotein particles that are secreted by the liver and intestine into the blood and serve to transport cholesterol through the system. The LDL particle consists of 25% protein in the form of a molecule known as apoprotein B, the rest of the mass being composed of lipid. Of the lipid portion, cholesterol constitutes 60%.

Because cholesterol is an essential component of mammalian cell membranes, cells cannot survive unless they synthesize cholesterol *de novo* or obtain it exogenously. When mammalian cells are grown in the presence of serum they will preferentially utilize the cholesterol available through the culture medium. This process takes place by way of the pathway shown in Fig. 13.1. The first step of this process involves the binding of the LDL to a surface receptor, most likely a protein. These LDL receptors are known to be localized in areas of the plasma membrane known as coated pits (Orci *et al.,* 1978). These coated pits contain 50 to 80% of the LDL receptors (Fig. 13.2).

The second step of the uptake process involves the internalization of LDL through endocytosis. Subsequent to this the LDL are absorbed in lysosomes and

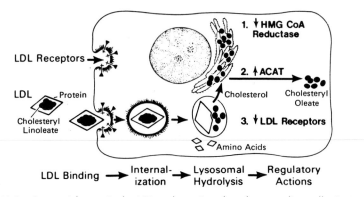

LDL Binding ➞ Internal-ization ➞ Lysosomal Hydrolysis ➞ Regulatory Actions

Fig. 13.1. Sequential steps in the LDL pathway in cultured mammalian cells. See text for explanation (Goldstein and Brown, 1979).

broken down into free cholesterol and amino acids. The free cholesterol is important in stabilizing internal cholesterol synthesis through a feedback mechanism. This is done by suppressing the synthesis of new cholesterol by affecting the activity of the rate-limiting enzyme 3-hydroxy-3-methylglutaryl-CoA reductase. Two other suppressive functions involve the packaging of cholesterol into esters through the activation of the enzyme acyl-CoA:cholesterol acetyl-transferase and the prevention of further entry of LDL into the cell by turning off the synthesis of LDL receptors. In this fashion the cell can regulate its internal supply of cholesterol in the face of constantly fluctuating external levels.

Mutations affecting the LDL receptor have been identified, and these are responsible for familial hypercholesterolemia. It is inherited as an autosomal dominant and is characterized by severe premature atherosclerosis and an elevation in plasma LDL. In homozygotes the manifestation is more severe, and death from coronary heart disease usually occurs before the age of 15 years. The elevation of cholesterol in familial hypercholesterolemia is entirely due to an increase in the number of LDL particles, an increase that results from the breakdown in cholesterol synthesis. The frequency of heterozygotes with familial hypercholesterolemia has been estimated at 1 in 500 (Goldstein *et al.*, 1973), making it among the most common of human genetic disorders.

Familial hypercholesterolemia has been studied *in vitro,* taking advantage of the fact that fibroblasts from affected individuals display the mutant phenotype in culture. To date, three classes of mutations have been identified on the basis of phenotype: (1) Rb^0, no detectable binding activity; (2) Rb^-, reduced binding activity; and (3) Rb^+,i^0, failure to internalize receptor-bound lipoprotein.

The first class, Rb^0, shows no high-affinity LDL binding. As a result of this, LDL is not internalized and HMG-CoA reductase and cholesterol synthesis are not suppressed; neither is cholesteryl ester formation activated. Cultured fibro-

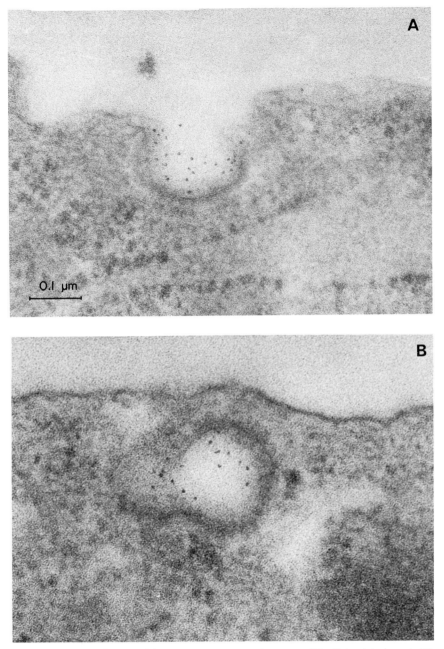

Fig. 13.2. Visualization of the LDL receptor system (courtesy of Dr. Richard Anderson). (A) The coated pit, lined with ferritin-bound LDL particles. (B) The internalization of such a vesicle.

blasts from homozygous individuals show total suppression for all features of the LDL pathway. However, when cholesterol is passed into the cell through other routes, suppression of HMG-CoA reductase occurs normally.

Rb^- alleles constitute the second group of mutations responsible for familial hypercholesterolemia. This is a heterogeneous class in which anywhere from 2 to 25% of normal receptor activity can be detected. This decrease in activity may be due to a decrease either in the number or in the affinity of LDL receptors. Individuals who appear to be compound heterozygotes have been detected. Such individuals have parents both of whom are heterozygotes, and the child inherits a different mutant allele from each parent in a fashion similar to individuals carrying different hemoglobin mutations, such as sickle cell and hemoglobin C (Hb^c/Hb^s).

A third type of mutation results in lack of the ability to internalize the LDL, apparently because of a failure to cluster the receptors in the coated pits. This mutation has been designated Rb^+,i^0 and has so far been observed in only one pedigree. The model formulated by Goldstein and Brown (1979) to account for this phenotype proposes that binding of the LDL occurs normally, but there is a failure of internalization, the result being that such individuals are unable to degrade LDL or utilize its cholesterol and hence display the secondary abnormalities of cholesterol metabolism observed in receptor-negative homozygotes. The failure of the receptors to cluster in the coated pits (Anderson et al., 1977) has been confirmed ultrastructurally. The authors favor the hypothesis that this mutation is a third allele at the same locus specifying the LDL receptor. The model for the assembly of the LDL receptor developed by Goldstein and Brown proposes that the LDL receptor molecule is a transmembrane protein that contains both a binding site for the LDL and a recognition site for coated pits. When functioning properly, the receptors will cluster on the coated pits and internalize LDL by pinching off to become coated vesicles. In the individual with the internalization defect, the receptors remain randomized on the cell surface and fail to internalize LDL into the coated pits.

The use of cell culture to isolate variants that are altered at different steps in the LDL pathway has been proposed. Goldstein and Brown (1979) have developed conditions in which cell growth is dependent on the LDL pathway. Cells with a mutation in the LDL pathway that prevents the uptake of LDL will fail to survive and can be isolated by negative selection (Chapter 3). In order for such a system to operate, it is necessary to completely shut off internal cholesterol synthesis, thus forcing dependency of the cell upon exogenous sources. This was accomplished through the use of a fungal antimetabolite that inhibits HMG-CoA reductase and that is known as compactin (Endo et al., 1976). Appropriate conditions for the use of compactin in cell culture have been determined, but mutants affected in the LDL pathway have not yet been isolated by this method.

A second approach involves the use of a reconstituted LDL particles containing molecules that are toxic to the cell. A cell that has lost the LDL pathway should be resistant to the toxic effects of such agents because of its inability to absorb them from the culture medium. Such preparations have raised the possibility that their toxicity could be used in the isolation of resistant mutants that are affected in the steps of the LDL pathway. Krieger *et al.* (1981) have used this method to isolate CHO cells that are receptor deficient. Using a reconstituted LDL particle containing toxic 25-hydroxycholesteryl oleate, surviving colonies were isolated. These were screened with a fluorescent LDL particle, and stable, receptor-deficient mutants were isolated.

A third approach utilizes cholesterol analogs. The oxygenated derivative 2,5-hydroxycholesterol causes an inhibition in the activity of HMG-CoA reductase. Mutants resistant to the toxic effects of 2,5-hydroxycholesterol have been isolated, and some of these variants have dominant mutations and therefore are constitutive for enzyme production. This suggests that regulation of this enzyme *in vivo* may involve an operator–repressor system (Sinensky *et al.*, 1980).

III. X CHROMOSOMAL INACTIVATION

In 1961, Lyon put forth an important hypothesis to explain the equal production of X-linked gene products in mammals. The proposal that one X chromosome is randomly inactivated in almost every cell of the female mammal is a theory with wide implications for the control of eukaryotic gene expression. Furthermore, it has served as a useful probe in the understanding of human genetics, malignancy, and cardiovascular disorders.

In the past it had been recognized that a mechanism must exist that would equalize the double dose of X-linked genes in females with the single dose in males. In *Drosophila,* Muller (1947) showed that this is accomplished by modulating the level of production of structural gene products from X-linked genes so that in a normal male there is twice the activity per gene of the normal female. This control mechanism is active irrespective of whether the genes are translocated onto an autosome; furthermore, autosomal genes translocated to the X are not modulated. Muller further showed that the ratio of X chromosomes to autosomes affected the level of expression, so that flies with three sets of autosomes and two sex chromosomes produced more activity on a per gene basis than did normal individuals. More recent observations (Lucchesi, 1978) suggest that the mechanism of dosage compensation in *Drosophila* results from autosomal compensator genes, which control the rate of mRNA transcription of the X-linked alleles (Belote and Lucchesi, 1980). These compensators may operate on groups of X-linked genes or on individual loci.

Because of the detail and the thoroughness by which this mechanism was described, it was difficult to conceive of an alternative system. In addition, the fact that the genetics of mammals was in a much more rudimentary state than *Drosophila* genetics made the accomplishments of Lyon (1961) and others all the more impressive. She proposed that the heteropycnotic, or late-labeling, X chromosome observed in the somatic cells of female mammals was functionally inactive. Furthermore, she argued that the inactivation occurred early in embryonic development and, in any cell, the functional X chromosome could be of either paternal or maternal origin. Originally it was asserted that the inactivation was irreversible and encompassed the entire X chromosome. The Lyon hypothesis was based on two observations: the normal phenotype of the XO female mouse and the dappled pattern of heterozygotes for sex-linked coat color mutations. Since that time a great deal of evidence has irrefutably established X chromosomal inactivation as the mechanism by which equal production of X-linked gene products in males and females is achieved. The major evidence is as follows:

1. Fibroblast cultures from women heterozygous for X-linked inborn errors of metabolism are a patchwork of normal and fully mutant cells and clones of such cells will perpetuate the normal or mutant phenotype. This has been demonstrated for many loci, including the genes for G6PD (Davidson *et al.*, 1963), HPRT (Migeon *et al.*, 1968), Hunter's syndrome (Danes and Bearn, 1967), and Fabry's disease (Romeo and Migeon, 1970).

2. Women heterozygous for color blindness possess a retina that is a mosaic of normal and color-defective cones (Born *et al.*, 1976).

3. A number of X-linked mutations affecting coat color in mammals (tabby in the mouse, tortoise in the cat) have a patchy phenotype in the heterozygous state (Lyon, 1961).

4. The X-linked disorder ocular albinism is characterized in the heterozygote by an optic fundis consisting of patches of pigmented and nonpigmented tissue (Franceschetti and Klein, 1954).

5. In the rare, sex-linked condition ectoderman dysplasia, heterozygous females are a mixture of patches of tissue with and without sweat glands (Passarge and Fries, 1973).

6. Testicular feminization (see Section V) is a defect in sexual development resulting from an absence of receptors for testosterone. Human females heterozygous for this mutation consist of a mosaic of cells with and without testosterone receptors.

The existence of X chromosomal inactivation has proved to be an extremely useful tool for the analysis of cell lineages in development and has yielded information bearing on two questions of fundamental significance: (1) the origin of tumor cells and (2) the formation of atherosclerotic plaques. With regard to the first question, the development of tumors in women heterozygous for glucose-6-phosphate dehydrogenase isozymes has been investigated. Whereas in samples of normal myometrium A and B type enzymes are found in nearly equal amounts, leiomyoma tumors always possess enzymes entirely of one type. Similarly, cells from individuals with chronic myelogenous leukemia have been examined for the presence of these isozymes, with similar findings (Fialkow *et al.*, 1965). These observations support the hypothesis that such tumors originate from single cells (Linder and Gartler, 1965).

The same approach has been applied to the analysis of atherosclerotic plaques with the result that such lesions also appear to have a monoclonal origin (Benditt, 1977). Earlier it was believed that such plaques, which can occlude the arterial passageway and cause the shutdown of the vessel, were a wound-healing reaction in response to irritation caused by cholesterol. However, more recent studies have determined that the interior of such lesions is composed of smooth muscle cells rather than fibroblastic scar tissue and could represent a benign tumor. This belief has found strong support from the observation that plaques obtained at autopsy from heterozygous $G6PD^a/G6PD^b$ women possess, in the vast majority of cases, only one form of the enzyme. Because tissue samples of a similar size from adjacent normal areas of the arterial wall contain equal mixtures of the two enzymes, the most likely hypothesis is that atherosclerotic plaques arise from single cells (Fig. 13.3). Thus, the pathogenic role of cholesterol may be to generate mutagenic epoxides which would act as carcinogens, rather than directly providing an irritant effect as previously supposed. Alternatively, cholesterol could serve as a transport vehicle to carry carcinogenic molecules into close proximity to the arterial wall. This may explain the correlation between cigarette smoking and heart disease, i.e., aryl hydrocarbons produced from cigarette smoke may be moved by this mechanism to the site of plaque production. There they may induce the enzyme aryl-hydrocarbon hydroxylase and convert the procarcinogens into carcinogens. This model of heart disease suggests a relationship to benign forms of cancer and explains many confusing observations on the etiology of this disorder.

Although the general concept of X chromosomal inactivation remains unassailable, two conditions that have been examined suggest that part of the X chromosome may not be subject to inactivation. Both the X-linked antigen Xg[a] (Fellous *et al.*, 1974) and the mutation responsible for congenital ichthyosis (sterol-sulfatase; Shapiro *et al.*, 1978) produce only clones with the dominant phenotype when heterozygous. This is the result that would be anticipated if both genes were functioning in each cell. In the case of sterol-sulfatase (EC 3.1.6.2), a lack of X chromosomal inactivation has been thoroughly established. Using mouse–human hybrids (Mohandas *et al.*, 1980), segregants were isolated that failed to express two human X-linked genes (*G6PD* and *PGK*) but did express sterol-sulfatase. Because the protocol by which the hybrids were obtained allowed the isolation of a cell with an inactivated human X, the sterol-sulfatase enzyme should not have been expressed unless part of X was not subject to X

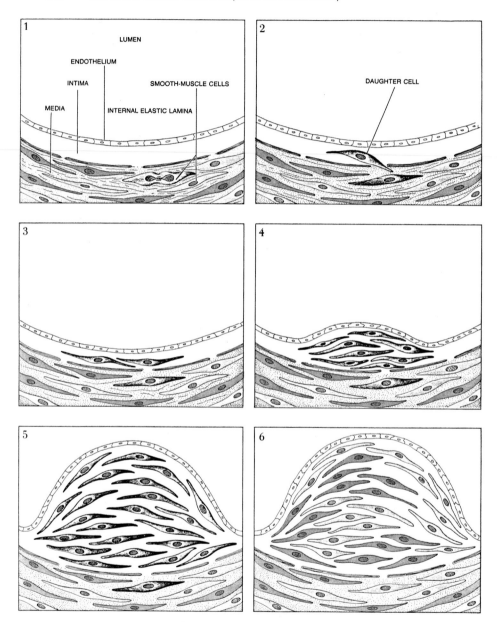

Fig. 13.3. Mechanism of atherosclerotic plaque formation according to the monoclonal hypothesis (from Benditt, 1977). Frames 1 through 5 show the progression by which a single cell undergoes a mutation and multiplies to form a small benign tumor. Frame 6 shows by contrast the appearance of a polyclonal plaque.

chromosomal inactivation (see Fig. 13.4). Both the sterol-sulfatase and the Xg^a locus are on the short arm of the X chromosome. When this region is absent, gonadal dysgenesis results. This observation is consistent with the proposal of Ferguson–Smith (1966) that a portion of the Xp arm is not subject to X inactivation, and the loss of this material results in the sterility observed in the XO female (Gordon and Ruddle, 1981).

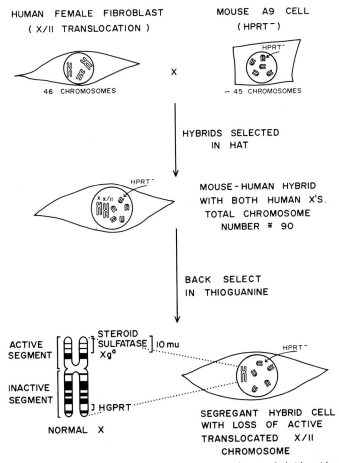

Fig. 13.4. Experimental protocol for obtaining mouse–human hybrids with a partially inactive human X. Human female fibroblasts with an X/11 translocation are fused with mouse HPRT⁻ A9 cells and selected in HAT medium. The hybrids are back-selected with thioguanine, and those cells with an active, X/11-translocated chromosome are eliminated. Remaining hybrids are those with a normal X containing an inactive portion and an active segment with the *SS* and *Xg*ᵃ genes.

Mohandas *et al.* (1981) have suggested a mechanism of X inactivation based upon DNA methylation. These authors have isolated a hybrid cell line with an inactive human X chromosome and a mouse X chromosome carrying a mutation for hypoxanthine phosphoribosyltransferase (see Chapter 2). Treatment of such hybrid cells with 5-azacytidine, which causes hypomethylation of DNA, results in a 1000-fold increase in clones able to grow in HAT medium. Analysis of individual clones showed that not all of the X-linked markers were activated, suggesting that X inactivation may involve segments of chromosomes rather than the entire X chromosome.

If DNA methylation is the mechanism of X inactivation, then how might the inactivation be stabilized through successive cell divisions? The authors suggest the existence of a methylase enzyme that carries out maintenance methylation on the DNA molecules, one strand of which is already methylated. Such a mechanism would explain the stability of X inactivation and the activation of genes by 5-azacytidine. These observations would seem to be in agreement with Kahan and DeMars (1975, 1980) who have studied reactivation of the human *HPRT* gene in hybrids between human female fibroblasts and mouse L cells. Activation of the *HPRT* gene of the inactive X was not correlated with the loss or gain of any specific chromosome, indicating that the continuing presence of a trans-acting regulator is not responsible for the Lyon effect. Activation of chromosomal segments of the X chromosome by hypomethylation would certainly be in accord with an autonomous mechanism of activation.

IV. XERODERMA PIGMENTOSUM

This extremely interesting genetic disorder is one of the best examples of the bridge between cell culture studies *in vitro* and investigations *in vivo* of a pathological condition. It is the result of a recessive mutation in a gene producing a DNA repair enzyme; affected individuals have an extreme sensitivity to ultraviolet light and a very high incidence of skin cancer (Cleaver and Bootsma, 1975). Affected individuals appear normal at birth but severe and progressive changes occur to the skin shortly afterward and are associated with exposure to uv light. These alterations are characterized by damage to the cornea, extreme freckling in areas of the skin exposed to the sun, marked atrophy of the dermis, and irregular proliferation of some of the layers. Multiple and progressive skin cancers occur in such large numbers that the patient usually dies before the age of 30. These tumors, both benign and malignant, are of both ectodermal and mesodermal origin and may be both basal cell and squamous cell epitheliomas (German, 1972). Cultured fibroblasts from such patients are defective in their ability to perform excision repair; specifically, such cells are deficient in the ability to excise thymine dimers, resulting in a reduction of the capacity to insert new bases. Cultured fibroblasts from individuals with xeroderma pigmentosum not only are much more sensitive to the killing effects of uv but also show an enhanced sensitivity to mutation, both as a function of uv dose and also as a function of survival. These results are in agreement with models of carcinogenesis proposing mutational mechanisms (Maher *et al.*, 1976).

Complementation tests demonstrate that at least five different complementation groups exist. This was shown by making cell hybrids in different pairwise combinations between fibroblasts taken from individuals with different forms of the disorder. If the mutational lesions are in different genes (cistrons), then the defects will mutually complement one another and will result in the formation of a functional product.

The isolation of mutants altered in their ability to repair damage to DNA would be especially useful in clarifying the genetics and biochemistry of xeroderma pigmentosum and related disorders, such as ataxia telangiectasia (Hoar and Sargent, 1976), Fanconi's anemia (Remsen and Cerutti, 1976), and Bloom's syndrome (Shiraishi and Sandberg, 1978). A number of reports have been published in which either replica plating methods or reverse selection methods were used to isolate variants sensitized to the killing effects of uv and mutagenic chemicals (for reviews, see Busch *et al.*, 1980; Thompson *et al.*, 1980). Thompson *et al.* (1980) reported a procedure for the isolation of such mutations in the CHO cell line; the procedure relies upon the recognition of damaged colonies at sublethal doses of mutagens. By visually screening large numbers of colonies and isolating and retesting those that appear to be especially sensitive, it is possible to isolate mutants that have no detectable uv repair and that appear to be similar to xeroderma pigmentosum cells. Busch *et al.* (1980) have employed a semiautomated technique, the "Cyclops/Microfilm Reader Method," for the large-scale isolation of uv-sensitive mutants. In this method large numbers of plates with small colonies are irradiated with sublethal doses of uv. The colonies are photographed, allowed to grow for 2 days, photographed again, and the colony sizes compared on a microfilm reader, which aligns the images slightly out of register. Using this approach, a large number of colonies with a defect in repair similar to xeroderma pigmentosum were isolated. The cyclops device may lend itself to the isolation of many types of rare variants.

V. TESTICULAR FEMINIZATION

Ordinarily, the presence of a Y chromosome in mammals causes the differentiation of the indifferent gonad into a testis. This is triggered by a masculinizing center on the Y chromosome which brings about testicular development, testosterone production, and the elaboration of secondary sexual development. However, this sequence of events is interrupted in XY individuals (humans, cattle, rats, and mice) that carry a sex-linked genetic disorder known as testicular feminization (Tfm; Ohno, 1971). This condition is characterized by a strikingly feminine external phenotype combined with a blind vagina and absence of internal genitalia except for testes. The testes are always internal, are degenerate, and do not carry out spermatogenesis.

The hormonal defect in testicular feminization is at the level of the androgen receptor. Although plasma testosterone levels are normal or higher than normal, there is a failure of cells in target tissues to respond, and male differentiation of these tissues is not brought about. The presence of circulating estrogens causes a subsequent development of female secondary sexual characteristics.

In mammals with the *Tfm* mutation there is a diminished binding of dihydrotestosterone (Griffin and Wilson, 1980). In humans there exist at least two mutations: (1) a thermosensitive mutation which results in normal binding at 26°C and decreased binding at 37°C, and (2) a mutation that causes a complete lack of binding at all temperatures.

Cell culture has proved to be a useful tool in the investigation of this defect because cultured fibroblasts possess testosterone receptors. Assays *in vitro* have shown that individuals with the syndrome have defective binding (Keenan *et al.*, 1974, 1975) and that, in agreement with the Lyon hypothesis, cultures from carriers are a mixture of cells expressing the normal and mutant phenotype. In keeping with the principle of conservation of the X chromosome, the *Tfm* mutation is sex-linked in all mammals. The mutation is recessive and carrier females are phenotypically normal.

The fact that mutants resistant to the toxic effects of steroids can be isolated in culture cells suggests that an *in vitro* model of the *Tfm* mutation and other, related forms of androgen resistance could be developed. If so, this might prove to be an extremely useful method of studying the biological and physical properties of not only the receptor system, but related steps as well, such as the enzymes involved in metabolizing androgens.

VI. CONCLUSIONS

The importance of cell culture in the development of human genetics has increased substantially in recent years, mainly as a result of the recognition of the molecular basis of a number of genetic disorders. One of the most notable aspects of this approach is the use of selective techniques for the isolation of variants that mimic human mutations. Because this enables the isolation of mutants not normally encountered *in vivo* (because of their rarity or lethality) and also because it offers the possibility of analyzing different mutants in cell hybrids, the next few years should see answers to a great number of questions related to the nature of various hereditary defects. Perhaps one of the most significant problems into which cell culture procedures could offer insight is the mechanism of action by which human genetic disorders produce particular phenotypic effects. For instance, the Lesch–Nyhan syndrome is known to be due to an absence of the enzyme HPRT (Chapter 2), but we do not yet understand how this results in the mental retardation associated with the condition nor what may be done to alleviate the behavioral symptoms, which are at present incurable. No doubt many of the manifestations of this condition result from regulatory alterations that could be analyzed *in vitro*. In addition, the induction of particular variants in culture and their reintroduction into a situation *in vivo* through the use of the teratocarcinoma system would be of great value (Chapter 10). It has already been shown that teratocarcinoma cells made drug resistant *in vitro* can be introduced into blastocysts and the particular variant can be established in a mosaic animal containing some cells with the induced mutation. Such an animal would contain sperm cells that were derived from the teratocarcinoma parent and that carried the mutation and could be used to establish a strain of (for instance) Lesch–Nyhan mice. This may prove to be a most successful answer to the analysis of human genetic disorders through the use of cell culture model systems.

14

The Cellular Basis of the Aging Process

I. INTRODUCTION

Most, but not all, multicellular organisms undergo a time-dependent loss of vigor that results in an increased probability of death. The regularity and predictability of this process is such that we can state, at any given point, the number of years that we can realistically expect to go on living (Comfort, 1979). The property of complex systems to run down and operate less efficiently with the passage of time is so universal that we take it for granted. Not only living creatures, but mechanical devices, governmental systems, societies, and even stars appear to have finite life spans. Because of the ubiquity of aging, it was accepted for much of human history as a natural and inescapable consequence of existence and the notion of understanding or modifying this process did not occur to most scholars, except in a frivolous fashion. As with other branches of scientific inquiry, the study of aging progressed from observational to experimental. Although early studies were based mainly on physiological considerations, in the past two decades specific hypotheses at the molecular level have been proposed to account for the mechanism of aging. In this chapter we will discuss the major hypotheses of aging and the experimental evidence that has been assembled in their support. As we will see, much contradictory evidence exists and a satisfactory synthesis to account for the many facets of this process has yet to emerge.

Before considering the cellular and molecular basis of this phenomenon, some general properties of senescence will be considered. Aging does not appear to be simply related to multicellularity of differentiation, because many plants (i.e., redwoods and bristlecone pines) appear to be essentially immortal and die only through happenstance. Similarly, observations on sea anemones indicate that they do not undergo senscence in an 80- to 90-year period (Strehler, 1962). Nor does size appear to be a factor in the aging process. Although certain large mammals (i.e., elephants, whales) appear to have long life spans, much smaller species (primates) reach comparable ages. In this regard, the best correlation between life span and size is the "cephalic index," a ratio of brain weight to longevity, which may have evolutionary implications (Fig. 14.1; Hayflick, 1980). Unfortunately, little accurate information on maximum life spans exists because, aside from laboratory species, much of such information is based on hearsay and exaggerated claims (especially in man) and does not appear to be reliable. It should be added that the study of human longevity is a fascinating, although frustrating, enterprise. This is gradually becoming based less on anec-dotal evidence and more on solid scientific data with the aid of recent improve-ments in techniques for dating biological materials. Thus, claims on extreme age among Russians living in Soviet Armenia have not been substantiated within the limits of error of the dating technique.

The average and maximum life spans for animals and plants in the wild differ tremendously (Comfort, 1979). In man there has been no increase in maximum life span since antiquity, whereas the average life span increased slowly from approximately 15 years during prehistoric periods up to approximately 40 years

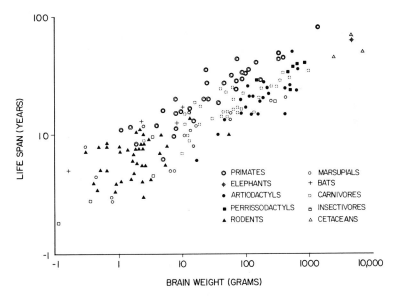

Fig. 14.1. Correlation between brain size and life span of various animal species. This suggests that there is a link between the two that arose over the course of the evolution of vertebrate animals. Data displayed in the graph were collected by George A. Sacher, Jr. of Argonne National Laboratory (Hayflick, 1980).

around the turn of the century. Since that time it has shown a rapid rise that is associated with the widespread use of public health measures throughout the industrialized world. However, at this point it is evident that the application to the general population of the most heroic medical advances will have little effect on increasing average longevity (Fig. 14.2).

The existence of a biological limit to the number of years that members of a species can complete suggests that aging is an orderly, determinate process, with its roots in a unitary mechanism. Foremost among the approaches to this problem has been the use of cell culture. Although other techniques have also been utilized, such as physiological studies of whole animals, the widespread use of *in vitro* approaches and the excitement it has generated make it a prime topic for consideration.

Hypotheses of cellular aging fall into two general categories. The first catego-

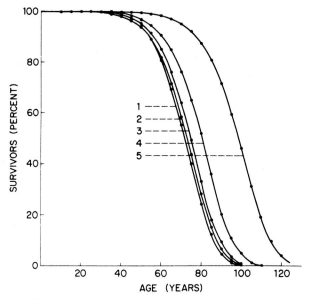

Fig. 14.2. Predicted power law plots of survivorship curves for humans, assuming cures of the three major causes of death. Curve 1 is for white males, United States, 1967 data. The median survival time is 71.8 years. Curve 2 assumes a cure of vascular lesions, yielding a median survival time of 73.0 years. Curve 3 assumes a cure for cancer, yielding a median survival time of 74.8 years. Curve 4 assumes a cure for heart diseases, yielding a median survival time of 80.5 years. Curve 5 assumes all three diseases are cured, yielding a median survival time of 98.8 years. It is implicit in these calculations that a cured individual does not have an enhanced susceptibility to any other cause of death. (From Rosenberg, 1973; reprinted in Comfort, 1979.)

ry are those hypotheses proposing that aging is due to changes within the nuclei of cells. Either these changes could be the result of a specific genetic program dictating cellular senescence or they could be due to random mutations accumulating over the course of time. A second category comprises those hypotheses that suggest that aging comes about as the result of an accumulation of detritus in the organism (cell, tissue, or organs) or as the result of nongenetic changes in the cell which eventually produce metabolic imbalance so grave that the organism can no longer function appropriately and death ensues. These hypotheses do not lend themselves easily to testing in cell culture and will not be considered in detail in this chapter.

One of the earliest hypotheses of aging is the somatic mutation theory (Hayflick, 1968), which in its most straightforward form suggests that aging is due to the accumulation of mutational damage over the course of time in cells of the body. When the level of mutational ''noise'' reaches a critical level, tissues and organs fail to function appropriately and death ensues. A more sophisticated refinement of this model is the ''autoaggressive'' hypothesis of aging developed by Burch (1979), which proposes that stochastic, mutational events occurring in central stem cells are required in order to initiate the aging process. Another generally discussed model of aging is the ''error catastrophe'' hypothesis (Orgel, 1963). In this hypothesis it is supposed that errors made by the cell's replicative machinery lead to the synthesis of faulty proteins, which in turn cause further synthetic errors at a rate that is exponential and eventually reaches catastrophic proportions—until the death of the cell occurs. These hypotheses have stimulated tests of their many specific predictions, and the testing has led numerous authors to endorse or reject the validity of the hypotheses.

The issue of the mechanism of cellular aging may be subdivided into several major questions:

1. Do primary cell strains (in particular, human diploid fibroblasts) possess a fixed, finite life span?
2. If so, is the process of cellular senescence *in vitro* the same as that which occurs in multicellular organism?
3. Are the predictions made by the error catastrophe hypothesis confirmed by experimental evidence?
4. If any particular specific proposal is found wanting, will alternatives such as the hypothesis of programmed aging account for the aging process?

II. THE AGING OF DIPLOID FIBROBLASTS *IN VITRO*

Hayflick (1965, 1979) has described a cycle of growth and proliferation that diploid fibroblasts undergo and that comprises three stages. Stage I represents the initial establishment of the cultures; stage II comprises reproducible growth through successive passages; and stage III involves progressive decreases in cell doublings and the eventual degeneration of the culture. Hayflick (1965) argued that such a process is not an artifact of the culturing procedure but rather represents a biological limit ("the Hayflick limit") imposed upon diploid fibroblasts as a reflection of normal aging.

If such is the case, then one would predict that the pattern of growth and division of cultured fibroblasts would fulfill a number of expectations. Particularly, one would expect that microbial contamination, culture conditions, type of medium, and method of transfer would not be implicated as causative mechanisms in the degenerations of cultures. These predictions appear to be reasonably well fulfilled.

In general, human embryonic tissues in culture possess a life span amounting to approximately 50 population doublings; however, there is extreme variability in this figure (Martin *et al.*, 1970; Smith and Hayflick, 1974; Schneider and Mitsui, 1976). The degeneration of diploid fibroblasts cannot result from depletion of the cell's own resources because in 50 cell divisions any preexisting molecules would have been long since diluted out. However, the possibility that aging is due to the loss of some critical molecular species which is synthesized more slowly than the cell replicates has not been completely eliminated. For instance, it may be that a protein molecule required for the initiation of DNA synthesis diffuses out into the medium more rapidly than it is produced through the metabolic activity of the cell.

Mycoplasma or other types of microbial contamination have been assiduously monitored and, although some degeneration of cells in culture may derive from this source, it will not serve as a general explanation for the limited replication of cultured fibroblasts. When cells with chromosomal markers (male and female) are mixed together in culture, the presence of aged cells does not shorten the life span of more recently initiated cultures beyond the normal expectation; nor will the converse hold (Hayflick, 1965). These results demonstrate that cell contact does not permit aged cells to transfer their senescent state to young cells and establish that infective agents are not responsible for the decline in the proliferative capacity of cultured cells.

The fact that cells can be stored indefinitely in liquid nitrogen and still retain their predicted number of cell divisions also indicates that a nonartifactual event related to the period that cells have spent in culture is responsible for their cessation of growth (Hayflick, 1968). However, these events are not strictly related to the number of divisions because, when cells are prevented from dividing by the depletion of serum in the medium or by agar overlays (Hay *et al.*, 1968), the cells will complete fewer total divisions. The number of divisions that they can complete is inversely related to the time that they have spent in culture in a nondividing state (McHale *et al.*, 1971). This is in contrast to the studies of Dell'Orco *et al.* (1973); these studies suggested that the number of divisions is a critical factor in the *in vitro* life span of serum-starved cells. Their studies indicated that the life span *in vitro* of cells maintained in a nondividing state for 77 and 177 days were lengthened by 91 and 210 days, respectively. The resolution of this controversy will require further experimentation.

Experiments using cell hybridization and cell enucleation techniques have been used to study the genetics of the decline in proliferative activity of fibroblasts. When diploid cells are fused with permanent cell lines they can be rescued from senescence and can be cultivated indefinitely. In mouse L cell–human diploid fibroblast hybrids, human chromosomes become a permanent part of the genetic repertoire of such cells (McKusick and Ruddle, 1977). This observation could be explained if aging were the result of recessive genetic changes in either the nucleus or the cytoplasm of the cell. Experiments by Wright and Hayflick (1975), involving the reassembly of different combinations of whole cells and cytoplasms enucleated with cytochalasin B, suggest that changes in the nucleus are responsible for degeneration because old cytoplast–young cell hybrids had the same proliferative capacity as young cytoplast–young cell hybrids. Furthermore, old cells fused with either young or old cytoplasts completed only a few divisions. A number of similar studies have led investigators to conclude that the cytoplasm does not play a role in cellular senescence (Bunn and Tarrant, 1980; Muggleton-Harris and Palumbo, 1979; Howell and Sager, 1978).

With regard to the nuclear contribution, experiments in which uncloned populations of diploid fibroblasts and permanent cell lines were fused suggest that unlimited growth *in vitro* is dominant in cell hybrids (Norwood *et al.*, 1975). However, analysis of clonal isolates (Bunn and Tarrant, 1980) demonstrates that most such hybrids have a finite life span, and escape from senescence in such clones is a rare phenomenon.

It could be that the finite life span of cultured diploid fibroblasts is the result of the accumulation of mutations in the cells brought about by the conditions of the culture system; this would predict that perhaps visible light (Jostes *et al.*, 1977) or nonphysiological concentrations of normal metabolites behave as toxic or mutagenic substances. The possibility that culture conditions could be responsible for an elevation in the mutation rate and the eventual cessation of the growth of such cultures through an error catastrophe has not been thoroughly explored. We have seen (Chapter 1) that the mutation rate in cultured cells, including both permanent cell lines and diploid fibroblast strains, is much higher than would be expected on the basis of the extrapolation of mutation rates *in vivo*. The influence of the *in vitro* environment could be tested by using the fluctuation test (Luria and Delbrück, 1943) to measure mutation rates in normal mouse fibroblasts obtained from embryos and by comparing these figures with mutation rate data obtained from normal mouse fibroblasts grown for a short period *in vitro*.

In sum, the evidence available at this time indicates that the limited proliferative capacity of diploid fibroblasts does not result from artifactual sources. However, certain possibilities have not been completely excluded, including the idea that culture conditions are causing an elevation in the somatic mutation rate, an elevation that results in the accumulation of lethal mutations and the eventual destruction of the culture.

III. PROLIFERATIVE CAPACITY OF DIPLOID FIBROBLASTS AND THE NORMAL AGING PROCESS

Although noteworthy in its own right as a biological phenomenon of some interest, the major impetus for the study of cellular division *in vitro* comes from its analogy to the natural aging process. If the analogy holds, then several questions must be answered in the affirmative.

1. Does donor age restrict the *in vitro* life span of diploid fibroblasts? Because the "Hayflick limit" of diploid fibroblast cultures *in vitro* is thought to result from a total utilization of a fixed number of cell divisions with which the individual is endowed, it follows that cultures obtained from aged donors should be able to complete fewer cell divisions *in vitro* than cells from younger donors. This prediction has been tested but the answer is not entirely resolved. Martin *et al.* (1970) obtained cultures from approximately 100 different donors and measured the number of cell doublings as a function of donor age. Those data demonstrated a decrease of 0.20 ± 0.05 cell doublings per year of donor age (Fig. 14.3). Although these observations have been frequently cited as evidence in support of the somatic mutation hypothesis, Kohn (1975) has called them into question, asserting that the variation is too great to calculate an accurate estimate of the effect of donor age on proliferation *in vitro*.

Schneider and Mitsui (1976) have studied *in vitro* life spans of fibroblasts derived from young and old donors, as well as migration rates of fibroblasts from skin explants, onset of culture senescence, and a number of other properties. Although these authors show that statistically significant differences exist between the two groups, inspection of the data indicates substantial overlap. It would appear that the great variability in the behavior of diploid fibroblasts *in vitro* may be a result of the variability in life span of members of the same species.

2. Do serially transplanted cell types *in vivo* show a decline in proliferative capacity related to the age of the donor? This question is still unresolved. Although a number of studies indicate that serially transplanted tissues from a variety of sources cannot be transplanted indefinitely from host to host, cells from aged donors do not appear to differ from cells from young donors in their capacity to proliferate when grafted into new hosts. Harrison (1975) observed that when bone marrow cells from normal animals were transplanted either into

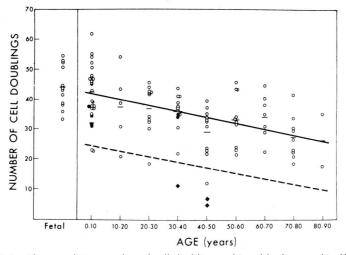

Fig. 14.3. The cumulative number of cell doublings achieved by human skin fibroblast cultures plotted as a function of the age of the donor. ○, Control cultures (see text); ♦, Werner's syndrome homozygotes; ▲, Hutchinson–Gilford syndrome heterozygote; ■, Rothmund's syndrome homozygote; ●, parallel cultures from the same donor (left- and right-arm biopsies from a 6-month-old girl and bisected, single biopsy from a 35-year-old man). The mean for each group (horizontal bar) and the calculated regression line (solid line) are shown (Martin *et al.,* 1970). (Martin *et al.,* 1970).

mice carrying the genetic anemia W/W^c or into lethally irradiated, normal genotype animals, there was no correlation between the age of the donor and its ability to restore functional marrow capacity.

Similarly, prostate (Franks, 1970), skin (Krohn, 1962), and mammary gland tissue (Daniel *et al.,* 1968) can be transplanted from animal to animal using isogenic donors. Prostate tissue could be serially transplanted for at least 6 years with no visible deterioration. Although the other two tissues eventually will lose proliferative capacity, skin survives much longer than the maximum life span of the host organism, whereas normal mammary gland lived no longer than 2 years (in isogenic mice). Thus, it would appear that the results of such serial transplantation experiments *in vivo* do not fit the expectations of the somatic mutation theory, at least in its simplest form.

3. Do physiological changes in cells cultured from aged individuals parallel the alterations seen in senescent fibroblasts *in vitro?* Studies using a variety of approaches answer this question in the negative. Schneider and Mitsui (1976) showed that modal cell volume, RNA content, and cellular protein content increase substantially in aging WI38 cultures. However, these differences did not exist between cultures taken from old as opposed to young individuals. Electron

microscopy of *in vitro-* versus *in vivo*-aged fibroblasts show no similarity in subcellular morphological degeneration (Robbins *et al.*, 1970).

So, the use of fibroblasts aged *in vitro* as a model for aging *in vivo* is open to serious criticism and cannot be considered yet as demonstrably valid.

IV. ERROR CATASTROPHE HYPOTHESIS

Orgel (1963) proposed that senscence results from the progressive accumulation of altered molecules with increasing age. At a certain point in time, errors would begin to appear in those molecules that are themselves responsible for the synthesis of other molecules. These errors would lead to other errors, resulting in an error catastrophe, with the loss of viability of the organism and eventual senescence. Since the time that this hypothesis has been put forth, a considerable body of literature aimed at establishing or refuting it has developed. There are a number of studies in which investigators have endeavored to establish that changes in proteins occur as a function of aging and that such changes are a result of a decrease in the fidelity of protein synthesis. Some of these studies have been reviewed and debated (Baird *et al.*, 1975; Holliday, 1975; Gershon and Gershon, 1976; Burch and Jackson, 1976).

Some enzymes have been demonstrated to increase in specific activity in crude homogenates during aging, whereas others were found to decrease in specific activity. Still others stayed the same. Holliday (1975) has argued that Orgel's (1963) hypothesis is consistent with these observations; Holliday asserts that those enzymes that show a decrease in activity do so as a result of faulty synthesis whereas those that show an increase in activity are synthesized in greater amounts as a result of a defect in a regulatory mechanism controlling their synthesis.

Perhaps the most important aspects of such studies relates to whether these changes are due to errors in translation or to other posttranslational mechanisms, such as addition of sugar moieties. Several authors have studied changes in glucose-6-phosphate dehydrogenase as a function of aging, either *in vivo* or *in vitro*. Holliday and Tarrant (1972) describe an increase in the fraction of heat-labile glucose-6-phosphate dehydrogenase in aging tissue and assume that such data support the error catastrophe hypothesis. However, these experiments were carried out using crude homogenates, and prudence must be employed when such data are evaluated. Rothstein (1975) has called into question experiments employing unpurified enzyme preparations, pointing out that the presence of stabilizing factors and proteolytic enzymes as well as the oxidation of sulfhydryl groups could alter the properties of the enzyme. Apropos of this point, Kahn *et al.* (1977) not only have shown that purified glucose-6-phosphate dehydrogenase from young and old fibroblasts shows no such difference in heat stability, but they also have shown that extracts from aged cells can cause alteration in the heat stability of glucose-6-phosphate dehydrogenase purified from human platelets.

In fact, Duncan *et al.* (1972) have observed that the accumulation of heat-labile glucose-6-phosphate dehydrogenase in aged fibroblasts appears to be due to a shift from the dimeric to the tetrameric form. Thus, the accumulation of the more heat sensitive enzyme molecules represents an equilibrium shift rather than a mutational change.

Lewis and Tarrant (1972) have published data that they assert are compatible with the error catastrophe hypothesis; these data show that aged cultures are less able to discriminate between ethionine and methionine in protein synthesis. But Rothstein (1975) pointed out that, on the contrary, their data show a constant, very low incorporation of ethionine into protein (probable background) together with a decreasing level of incorporation of methionine, reflecting a lower capacity for protein synthesis in the cultures that have reached senescence.

Age-dependent changes in enzymes of macromolecular synthesis have been found as predicted by the error hypothesis. DNA polymerases from mouse liver (Barton *et al.*, 1975) and from human fibroblasts (Linn *et al.*, 1976) decrease in fidelity with age. Concomitantly, there is a decrease in the rate of DNA repair (Epstein *et al.*, 1974; Mattern and Cerutti, 1975). It remains to be seen whether these alterations are genetic, are due to posttranslational modification, or are the result of faulty synthesis.

Fulder and Holliday (1975) have employed a different approach to the question. Taking advantage of a histochemical stain for glucose-6-phosphate dehydrogenase, they have measured the frequency of cells that can utilize 2-deoxyglucose-6-phosphate as a substrate for the enzyme. They have found an exponential increase in the frequency of staining cells in late-passage cultures, an observation that they claim supports Orgel's (1963) hypothesis. Several questions remain unanswered in this interesting work; among them is the genetic nature of the variants in question. It is possible that the change is due to an induction of nonspecific monophosphate hexose dehydrogenase in late-passage cells and that the age-related defect is in the regulation of enzyme production. The authors offered indirect proof that the variants are indeed mutants, although, as they point out, isolation and subculturing would be required to definitively establish this issue. Another problem regarding this work relates to the previously discussed question of posttranslational modification. Until it can be proved (by mapping of the mutant gene or by detailed chemical analysis) that the variants represent mutations within the glucose-6-phosphate dehydrogenase structural locus (located on the X chromosome), such studies will remain problematical.

Other types of experiments have produced findings that do not seem to be in agreement with the error hypothesis. One is the observation that diploid cells grown in sublethal concentrations of amino acid analogs do not have a shortened life span (Ryan *et al.*, 1974). If aging were due to mistranslation in protein synthesis, it would be expected that agents that expedite this process would shorten the life span of cultured cells. The other consideration is the fact that viruses grown in old as opposed to young fibroblasts are not altered in specific infectivity, kinetics of thermal inactivation, or mutation rate of virus-induced replicase (Holland *et al.*, 1973; Tomkins *et al.*, 1974). Also, the fact that tumor cell lines can be maintained through an indefinite number of passages argues against the error hypothesis.

It seems that if aging results from the breakdown in the regulation of some essential and limiting process, then the evidence supporting the error hypothesis, based on alterations in enzyme activity, may be fortuitous. The loss of regulatory control could be caused by any one of several mechanisms, including errors in the synthesis of macromolecules, mutations in regulatory genes, or changes in gene expression resulting from a specific differentiated program. Even though the rate of misincorporation of amino acids into polypeptides has been measured (Loftfield, 1963; Loftfield and Vanderjagt, 1972), the change in this rate with advancing age has never been shown. This observation would indicate that errors in translation of proteins are not the primary cause of senescence.

Two recent studies argue very strongly against the error castrophe hypothesis. Gallant and Palmer (1979) have argued that it will not satisfactorily account for the aging process because even a tenfold amplification of translational error induced by streptomycin in bacterial cells will not bring about an error catastrophe. Their consideration of the fidelity of translation in somatic cells leads them to conclude that higher organisms are no more subject to error propagation than bacteria. This conclusion is reinforced by Gupta (1980), who demonstrated that repeated cycles of mutagen treatment fail to shorten the life span of cultured fibroblasts.

V. CONCLUSIONS

The models under consideration at this time have serious problems; one of the gravest objections to the mutational models of aging lies in the quantitative discrepancy between requisite and actual mutation rates in somatic cells. If the average mutation rate is 10^{-6} per cell per division (Chapter 1), if there are 2×10^4 diploid loci whose inactivation could result in cell death (Comfort, 1979, p. 217), and if the mutations are recessive, then the probability of a lethal mutation that will inactivate the cell is

$$P_{\mathrm{L}} = 10^{-6} \times 10^{-6} \times 2 \times 10^4 \times 10^2 = 2 \times 10^{-6}$$

That is, only two cells per million will be inactivated after 100 cell divisions. If the mutation rate in normal somatic cells is as low as suggested (Chapter 1;

Morrow, 1977), then the figure would be even lower. This is a serious objection in the absence of some error-amplification mechanism. However, we have seen that the best described error-amplification model (i.e., error catastrophe) has serious problems as well.

One other hypothesis may bear renewed scrutiny. Aging has been described as terminal differentiation (Hayflick, 1973), and indeed it possesses an ordered component that is reminiscent of the differentiation process. According to this hypothesis, specific genetic systems would activate degenerative processes at a programmed point in the life cycle of a particular given species. These degenerative changes would be tissue specific and would reflect the particular differentiation program of the organ in which they occurred. As such they would be genetically determined and would be the result of evolution and natural selection for a biological mechanism that functions optimally at a period when mortality is at its nadir and vigor at its zenith. Thus, aging would result from some developmental processes that are highly advantageous during early life but that have delayed harmful side effects that eventually lead to senescence and death (Strehler, 1979). The virtue of this model over the somatic mutation hypothesis of aging is the fact that it does not call for a random mutational mechanism dependent on the total number of cell divisions. Not only would such a hypothesis explain the fact that different species have different life spans, but it also would explain why malignancy allows an escape into immortality in the same fashion that tumor cells so frequently lose the differentiated features of their tissues of origin.

This model would predict the existence of a limited number of loci responsible for aging and would suggest that they could be identified in cell hybrids between normal diploid fibroblasts and permanent cell lines. However, the experiments of Bunn and Tarrant (1980) indicate that senescence is dominant. So, if aging is due to particular genetic factors that are dominant in cell hybrids, one would expect to see rare segregants in which the ''aging genes'' had been lost and that had become immortal. This, in fact, may be what is occurring; and this hypothesis will be supported if the mouse–human hybrids that have been so widely studied (McKusick and Ruddle, 1977) can be shown to have consistently lost particular regions of their human genome.

Finally, a completely different, and iconoclastic, approach to the study of aging should be considered. Rosen (1978) has reviewed the literature on aging and has concluded that the reductionist approach to the study of aging may not be satisfactory for understanding a system that he believes to be composed of interrelated networks. He asserts that in many complex systems apparently minor changes in individual components can result in drastic changes in the state of the overall system. Thus, a trivial change in frequencies in the elements of the population can force the population into an entirely new state. Because most studies on aging have dealt with the properties of individual cells, it is quite possible that these investigations will miss the network aspects of the system which Rosen believes are responsible for its failure. This is a rather disturbing assessment because it suggests that the time-tested approaches to experimental biology may not resolve the mechanism of aging. What may be required is "a set of concepts and techniques totally different from those currently being employed."

We therefore find ourselves some distance from a resolution of the aging question. However, the existence of specific molecular models of aging (for instance, the role of transpositions; see Chapter 15) and the experimental tools to examine them should bring us to a much more profound understanding of this phenomenon within the next decade.

15

Future Outlook

I. INTRODUCTION

The tremendous advances made in the field of cell genetics in the past 20 years are all the more exciting when one considers their implications for future discoveries. The technical developments described in this volume, together with the unfolding of our theoretical understanding, have immensely improved our knowledge of cellular biology. However, despite these developments there are wide gaps in this theoretical framework, and about some areas, such as differentiation, we are largely ignorant.

In this chapter we will consider our overall state of knowledge of the field and its future direction. Although there is a certain risk inherent in prediction, it serves to focus and to develop our concepts for research in these areas.

One generalization that can be made concerning the future of cell genetics is that it will be more and more closely related to human welfare and practical technological applications. This will mean that biological scientists will become more concerned with technical, legal, and financial considerations than at any previous time. Although the interdependence of science and technology upon political and financial establishments is probably as old as governmental institutions, this is the first time that molecular biology has participated in such an intimate involvement. It is certain that not all the fruits of this relationship will be positive; every week a new litigation or financial debacle is reported in the "News" section of *Science* (see, for instance, Wade, 1980). Thus, the topic to be considered in this chapter possesses a cutting edge of more than academic significance.

II. RECOMBINANT DNA TECHNOLOGY

The use of endonucleases for the analysis of eukaryotic gene structure has proceeded at an electrifying pace—so rapidly that data on sequences is accumulating more rapidly than is our understanding of them. However, the continuing analysis of genes will certainly produce repeatable patterns that should resolve a number of questions. Among these is the nature and function of the intervening sequences (introns). Are these sites especially liable to recombination, breakage, pairing, or crossing over? Are such regions somehow involved in the regulation of the rate of transcription? Chromosomal structure in higher organisms is tortuous and based upon patterns of coiling and supercoiling; perhaps these intron regions are areas that allow for looping and pairing with other parts of the chromosomal strand and for the development of a three-dimensional order.

Not only is the recombinant DNA technology of great value to the investigation of eukaryotic chromosomal organization, it is also of use in the construction of molecules that carry genetic information. The possibility that substances of great biological significance, such as interferon, insulin, and human growth hormone, could be synthesized through the use of recombinant DNA technology has excited tremendous interest because the economic and medical potential of such techniques is substantial. Some caution might be wise at this point, because recombinant DNA technology has yet to generate substantial profit and because many of these substances may in the long run be more easily produced through less exotic approaches such as adaptation of lymphocytes for large-scale production of interferon *in vitro*.

Other innovative applications of biotechnology have stimulated a great deal of popular interest. For instance, the notion that a strain of bacteria could be developed through ordinary selection procedures or through genetic engineering to break down hydrocarbons has been mentioned as a way of combating oil spills. It has also been asserted that this might represent a doomsday machine for the petrochemical industry because if such an organism escaped it could degrade all the petroleum in the world and leave us without energy (and presumably leave the OPEC countries without working capital). This whole discussion has an air of unreality. A bacterial strain with the ability to divide in an unlimited fashion under the extremely adverse conditions of temperature and nutrition present naturally in the environment would be extraordinary. If this were not the case, a bacterium with the potential for eliminating all the petroleum in the world would no doubt have already arisen through evolution and natural selection, and at this stage our planet would be devoid of this substance.

In fact, nature seems to have a way of frustrating biological doomsday epics which tend to make the initial fears of writers and many scientists on the horrors of genetic engineering seem ludicrous. Thomas (1977) reviewed the debate on genetic transfer experiments and concluded that much of it was farfetched and quite unrealistic. His conclusions, drawn several years ago, appear to be more and more accurate as information on the actual, as opposed to the imagined, hazards of genetic engineering accumulates. In addition, he presents a highly critical evaluation of arguments concerning hazards arising from the accidental release of new, genetically engineered strains of microorganisms into the environment. Nor does he feel that genetic engineering offers much promise for the alleviation of human genetic disorders. Thomas in this essay goes so far as to state that ''talk about genetic engineering and the manipulation of human genes is . . . irresponsible.''

Although the essence of Thomas' remarks is accurate, he may have overstated his case when he argued that genetic engineering for the treatment of genetic disorders is so fraught with problems as to be completely impractical. The transfer of the tetrahydrofolate dehydrogenase gene into mouse bone marrow cells proves that certain genetic engineering techniques are practical for the insertion of genetic information into whole organisms (Cline *et al.*, 1980; Chapter 2). In particular, treatment of diseases that do not require a complex organization of tissue for a cure to be effected are of particular promise. These include many genetic disorders of the blood, such as hereditary anemias and hemophilias. Thus, it is possible that defective genes could be treated by removing the host cells, inserting the appropriate genetic information into them, and replacing them.

This approach has already been used in humans in an attempt to treat the disease β^0-thalassemia, in which no hemoglobin β chain is produced (as reported by Kolata and Wade, 1980). This is a fatal disorder, and its progress is easier to monitor than is that of sickle cell anemia, for which this treatment might serve as a model. It is not yet clear whether this highly controversial strategy will be successful.

Although not involving the use of gene-splicing techniques, one of the most striking treatments of pathological conditions through the injection of special cell types is the implantation of neurons into the brain. Perlow *et al.* (1979) have shown that in an animal model of Parkinson's disease it is possible to achieve marked improvement of motor function through the injection of fetal neurons into the caudate neucleus of the brain. These studies showed that grafting of neurons from fetal ventral mesencephalonic areas produced a repopulation of the denervated area and a marked restoration of normal behavior. Work by another group (Stenevi *et al.*, 1976) showed that several types of fetal neurons could be implanted into the adult brain of the rat and survive consistently, provided they were in contact with an appropriate vascular bed. A more recent report from the same laboratory (Bjorklund and Stenevi, 1979) shows that lesions produced in the nigrostriatal pathway by 6-hydroxydopamine (6OHD) can be reestablished using embryonic neurons from the same area. Furthermore, motor lesions induced by the 6OHD treatment are to a large extent alleviated following this procedure.

These findings suggest that in the future genetic or environmentally caused disorders of the brain might be cured in human beings through injection of neurons, either neurons treated with DNA developed by recombinant DNA technology or simply neurons from a donor with normal function. These neurons could be maintained in culture, and the fact that the brain is an immunologically privileged site (Shuttlesworth, 1972) makes such experiments all the more feasible.

A related area of research, although not quite as full of promise as the manipulation of human genes, is genetic engineering in plants. There has been no shortage of ideas put forward to improve plant strains, such as insertion of nitrogen-fixing genes by transformation or transduction into crop plants (especially monocots). Despite a very substantial effort in this area, results to date have not been particularly gratifying. This failure is due both to the extremely high energy requirements of nitrogen fixation as well as to its anaerobic nature. In a similar fashion, the use of plant somatic cell culture has been proposed as a means whereby large populations of individuals could be subjected to selection and mutagenesis in order to obtain useful variants that could then be grown into complete plants. This strategy has been successful in developing strains of tobacco resistant to methionine sulfoximine, which is an analog of the bacterial toxin that is responsible for wildfire disease (Carlson, 1973). The isolation of strains of plants that are resistant to toxins produced by bacteria or fungi through the use of somatic cell genetics may be of value. It is difficult to see, however, how such a technique could be used in the vast majority of cases. Many plant diseases involve complex patterns of interaction between more than one vector, and they exert their toxicity through mechanisms that do not involve killing by antimetabolites.

Another area of plant genetic engineering that has not lived up to its initial high expectations is the area of somatic cell hybridization using plant protoplasts. Viable fusion products can be obtained *in vitro* and entire hybrids can be cultivated from protoplasts. Plant cell hybrids have been generated with two species of tobacco; this process yields, through somatic cell fusion, an artificial hybrid with all the properties of a sexually produced hybrid (Smith *et al.*, 1976). However, with this approach no hybrids have been generated that were not already available through normal sexual crossing. Furthermore, protoplasts must be produced through enzymatic treatment and are much more fragile and difficult to handle than tissue culture cells.

Cocking *et al.* (1981) have reviewed the present state of genetic manipulation in plants and they cite a number of important difficulties that remain to be surmounted before these techniques can be applied to agricultural practice.

1. There is a paucity of stable auxotrophs in plant cells; thus, the development of transformation and other gene transfer systems is difficult.

2. Many of the characters in plants that are associated with increased yield are controlled polygenically; it is unlikely that such collections of genes could be transferred as a unit.

3. Although it appears to be relatively simple to bring together two plastid genotypes in the same cytoplasm, it has proved difficult to keep them together.

4. Progress in the application of somatic cell genetics to plants has been impeded by the fact that relatively few genera can be regenerated from protoplasts; those that can be regenerated are of limited economic value.

5. Fusion of protoplasts from widely divergent species is followed by the elimination of chromosomes and the loss of valuable genetic information.

6. Nitrogen fixation has substantial energy requirements in addition to its anaerobic requirements; thus, the technical problems involved in the introduction of this process into plant cells, especially those with economic value, are substantially increased.

At this point, the actual accomplishments of research in plant genetic engineering and somatic cell genetics are somewhat limited. This is not to say that impressive developments in this area do not lie in the future, but merely to indicate that scientific expectations and real discoveries may not mesh too closely. Butenko (1979) has reviewed the use of plant protoplasts in genetic transfer studies, including somatic cell hybridization and DNA transformation. Although he is optimistic concerning the future of research in this area, it appears that many difficulties remain to be resolved. At this time, there is no thoroughly substantiated case of a transfer of genetic information utilizing DNA, phages, phytoviruses, or plasmids in plants.

III. IMMUNOGENETICS

Consideration of basic research efforts in this discipline gives a more encouraging picture. Some of the most impressive scientific advances have been made in immunology, in which the use of recombinant DNA technology has been brought to bear on the precise mechanism of antibody synthesis. Perhaps the most revolutionary and dramatic finding to come from the study of immunology has been the realization that differentiation of the immune system proceeds through a slicing and rearrangement mechanism, in which segments of genetic information are transposed in an orderly manner so that they come to lie in close juxtaposition to one another. This finding may have extremely broad implications for eukaryotic gene expression. As mentioned earlier, transposition of blocks of genetic material may be crucial in the etiology of malignancy (Cairns, 1981). Furthermore, the control of differentiation in general may lie in a mechanism by which genes are shuffled throughout the genome. Finally, the basis of cellular aging may be related to gene transposition. It appears that within the next decade most of the major issues regarding the means by which antibodies are formed and the various cellular responses associated with the immune system are generated will be resolved. Although a thorough understanding of a scientific phenomenon is not necessarily synonymous with its control (for instance, Down's syndrome is completely understood but cannot be cured or controlled), in this case it is clear that the treatment of a variety of disorders with tissue grafts from unrelated donors will be made possible by an understanding of the mechanism of antibody formation.

IV. HUMAN GENETICS

The examples discussed in Chapter 13 point out a high level of sophistication in areas of human genetic research in which specific proteins have been identified and their role in disease processes described. However, there are vast areas of total ignorance in the field of human heredity. The most fundamental questions that the average person has about genetics—"Why does my child resemble me?"—cannot be answered with any degree of intellectual rigor or detail. This is due to our limited understanding of multifactorial inheritance and to a lack of information on regulatory systems in higher organisms. It is likely that many of the approaches described in this volume will not be particularly useful in resolving these questions, because complex developmental systems requiring a particular framework of space and time could not be studied in isolated cell culture systems.

Obviously, the place to start in the analysis of developmental regulation would be through the use of mutants affecting the process in clearly obvious ways. This line of investigation is still a long way from the molecular level because at present we lack the ability to identify the gene product for many dominant disorders, especially those causing developmental abnormalities. It is not even clear whether these genes are involved in the production of enzymes, regulatory proteins, or simply nontranslated RNAs that regulate other genes.

However, cell genetic approaches have figured prominently in the development of our understanding of single-gene defects in man (Chapter 13). Over the course of the next few years we can anticipate that the pathology of a number of genetic disorders will be solved with the aid of cell culture. For instance, the Lesch–Nyhan syndrome is thoroughly understood at a biochemical level; however, we still have no notion as to the cause of the mental retardation that accompanies it. Probably one of the most exciting strategies for this purpose is the use of the teratocarcinoma system combined with *in vitro*-generated cellular variants. It is now possible to obtain cell variants in the teratocarcinoma cell line, insert them into blastocysts, and produce mice with specific genetic disorders that mimic the human disease condition; this has already been done in the case of Lesch–Nyhan syndrome and thioguanine resistance.

Cystic fibrosis is another condition whose analysis has proved to be extremely frustrating. Despite a tremendous research effort, we still do not know the primary gene product of this disorder, and only preliminary clues as to its nature exist at this juncture. The search for an understanding of this disease at a cellular level is one of the most pressing tasks of somatic cell genetics. One approach to this problem would be to work backward from the gene to the product. This might be accomplished by screening mutants for populations of mRNA molecules present in the wild type but not in the mutant. The feasibility of this approach has already been demonstrated in yeast, in which it is possible through cross-hybridization procedures to identify the mRNA for the galactose-inducible genes (St. John and Davis, 1979). This was done by producing cDNAs from induced and uninduced cells and cross-hybridizing them. From such procedures unique sequences were isolated from the induced cells but failed to hybridize with the cDNA of the uninduced cells. Applying this procedure to particular genetic disorders could serve as a probe for the identification of the particular genes. This technique might be applied to cystic fibrosis in order to isolate particular genes whose messengers are not produced in the mutant. With such probes the gene products might be identified, the altered genes might be localized to chromosomes, and the possibility of prenatal diagnosis through genetic probes could be explored.

Recombinant DNA technology will be utilized more and more in the coming years to diagnose genetic disorders prenatally. Many disorders result from mutations that affect differentiated gene functions and whose gene products are, therefore, not present in amniotic cells. When DNA probes are developed for genes such as those responsible for phenylketonuria, albinism, and hemophilia, it should be possible to diagnose such disorders rapidly and safely from amniotic cells, as is already being done for thalassemia and sickle cell anemia.

V. AGING

Research in aging has become bogged down in efforts to establish or refute various theories of cellular senescence and to relate these to organismal aging. Large amounts of data continue to be accumulated, but it is difficult to see how real progress will be made until a new theoretical framework allowing research to move forward on different lines is developed. The failure to establish a unitary mechanism for the aging process may be due to a single organism. This possibility, however, does not alter the value of studies on the theoretical basis of aging. If the many competing theories of aging can be eliminated through truly decisive experiments or if completely new models of aging can be developed, real progress may occur.

One very interesting hypothesis to explain aging has been proposed by Macieira-Coelho (1980). He suggests that aging may be the result of the movement of transposable elements from one part of the chromosome to another and of a subsequent inactivation of needed functions. Although there is no evidence available to support such a model, it fits in well with other notions mentioned here concerning the role of transpositional changes in cancer, development, and immunology. Furthermore, the variability observed in cultured cells (Chapters 1 and 2) points to the possible importance of transposition in this phenomenon. The testing of this model should be possible in the near future by using recombinant DNA technology.

VI. MALIGNANCY

This area may be the most likely in which to expect significant developments to be made both on a theoretical and a practical level in the next few years. Cancer research has received such a tremendous expenditure in terms of time, money, and talent that it would be surprising if impressive gains were not forthcoming.

Cairns (1978), however, has formed a rather pessimistic assessment of the immediate prospects for breakthroughs in either the treatment or the theoretical understanding of cancer, and he is not alone in this conclusion. His argument is essentially that real progress in the treatment of human disease has been part of a slow, gradual continuum in human history and that there is no reason to believe that cancer research will prove to be any different. He points out that major strides in the treatment of viral disease, including vaccination, which has a history extending back to Egyptian times, were based on slow, progressive increments in our understanding. He further argues that because neither treatment, diagnosis, nor prevention appears to be particularly effective at present in reducing cancer mortality, we have no basis for optimism about developments in the near future, and that a cure for cancer might not be realized until the next century.

Although Cairns's position is astutely reasoned, there may be flaws in his argument, and it is realistic to hope for a much shortened timetable. Although cancer is an extremely intractable disease that will not yield to any single standard approach such as using antibiotics ("wonder drugs"), there are various tantalizing prospects on the horizon. Although clinical trials with interferon have so far been mixed, there is certainly reason to believe inroads into cancer mortality can be made using this substance.

On the other hand, the existence of a number of biological growth-promoting substances suggests that it may be possible to develop other approaches to the control of cellular growth in its abnormal states, approaches that do not depend upon devastating chemotherapeutic treatments.

Another mechanism for the treatment of cancer that Cairns has not considered is the use of monoclonal antibodies. There are a number of mechanisms by which these substances might be of use in the control of cancer. McGrath *et al.* (1980) have published results of studies in which monoclonal antibodies were isolated against mouse lymphoma cells and were tested for their ability to block virus binding and for cytotoxic activity. The fact that those antibodies that possess one property also possess the other suggests that such cell surface sites are receptors involved in growth promotion. This suggests that cancer cells might be inhibited with some degree of specificity. Another means of treating neoplastic cells by utilizing the hybridoma system entails the isolation of monoclonal antibodies to tumor-specific antigens, followed by the coupling of these substances to toxic compounds such as diphtheria toxin. Such substances could act as specific "chemical arrows," which would only be active against cancer cells. This approach has been attempted by Gilliland *et al.* (1980), with encouraging results. These authors have prepared a monoclonal antibody against human colorectal carcinoma cells and coupled this to the A chain of diphtheria toxin. These preparations appear to be specific inhibitors of protein synthesis in cultured carcinoma cells, although their specificity *in vivo* has not yet been determined.

It is noteworthy that these approaches do not assume any special knowledge about the basis of neoplasia that we do not already possess. We hope that such information will be forthcoming and will add to a rational group of therapeutic treatments against cancer. There are a variety of models that have been proposed to account for malignant growth. Elements of many of these theories may be correct, and these represent many possibilities by which neoplastic disorders may be attacked.

VII. DIFFERENTIATION

This question is related to neoplastic disorders in a general sense, and its resolution represents one of the major challenges of modern biology. Because genes controlling differentiation do not produce a product that we can identify at present, the most promising approach may be the use of restriction endonucleases and DNA sequencing. Although we do not know the function of the introns that lie within structural genes, it is possible that they are responsible for important regulatory roles in multicellular organisms. An analysis of developmental mutants, such as the T alleles in the mouse, could produce important insight into the role of intervening segments in DNA in the regulation of differentiated functions.

Another approach to the question of differentiation that has stimulated interest recently is the possible role of DNA methylation. Because there are no restriction enzymes in eukaryotic organisms, the methylation of DNA must serve a different function from that which it performs in prokaryotes. This role may be a regulatory one, for it is known that the incorporation of the methyl group of 5-methylcytosine into the major groove of DNA affects the binding of regulatory proteins (Riggs, 1975).

Furthermore, these modifications could be clonally inherited. It is known that DNA methylases act rapidly to methylate a half-methylated site. Indeed, after methylation the deamination of 5-methylcytosine could result in a directed mutation (McGhee and Ginder, 1979). There has been some exploration of the methylation of differentiated genes; however, a rigorous analysis of the role of this phenomenon lies in the future.

VIII. CONCLUSIONS

In discussions of the history of molecular biology and genetics there has been talk on numerous occasions of a "golden age" of molecular genetics. It is suggested that during this period, principally in the 1950s, all the foundations of our molecular understanding of living systems were formulated (Sibatani, 1981). According to this historical view of molecular biology, the subsequent work represents a "mopping up," during which the earlier findings were consolidated. Sibatani has argued that the property that distinguished the "golden age" from later periods was the fact that an intellectual framework existed that predicted certain discoveries (such as messenger RNA) before they were actually made. According to this interpretation, the present thrust of activity is not a second golden age, because none of the discoveries were predicted by a heuristic intellectual framework; nor do these investigations, brilliant as they are, go to the depths of the theory of life.

If this interpretation is correct, then do we have reasonable hopes for a true "second golden age"? I believe that the answer to this question is "Yes." Up to this point, the development of our understanding of complex biological processes has relied upon the use of analytical processes by which these systems are understood through dissection of their component parts. It may be that certain complex phenomena such as aging, malignancy, and differentiation cannot be understood simply through an analysis of their component parts and that it will be necessary to develop new approaches that are integrative or global. It is not at all clear how such interacting systems might function or what their bases might be. The difficulty in formulating a model of an interrelating biological system makes it clear how substantial a task its experimental resolution would pose. If these problems require the use of new approaches, it will represent the most exciting challenge that modern biology has yet provided.

Bibliography

Adair, M., Thompson, L. H., and Fong, S. (1979). [^{3}H]Amino acid selection of aminoacyl-t-RNA synthetase mutants of CHO cells: Evidence of homo- vs. hemizygosity at specific loci. *Somatic Cell Genet.* **5,** 329–394.

Aebersold, P., and Burki, H. (1976). 5-Bromodeoxyuridine mutagenesis in synchronous hamster cells. *Mutat. Res.* **40,** 63–66.

Affara, N., Robert, B., Jacquet, M., Buckingham, M., and Gros, F. (1980). Changes in gene expression during myogenic differentiation. *J. Mol. Biol.* **140,** 441–470.

Albertini, R. J., and DeMars, R. (1973). Somatic cell mutation detection and quantification of x-ray induced mutation in cultured diploid human fibroblasts. *Mutat. Res.* **18,** 199–224.

Alt, F. W., Kealems, R. E., Bertino, J. R., and Schimke, R. T. (1978). Selective multiplication of dihydrofolate reductase genes in methotrexate-resistant variants of cultured murine cells. *J. Biol. Chem.* **253,** 1357–1370.

Alter, B., Goff, S., Hillman, D., Deisseroth, A., and Forget, B. (1977). Production of mouse globin in heterokaryons of mouse erythroleukaemia cells and human fibroblasts. *J. Cell Sci.* **26,** 347–357.

Amano, T., Hamprecht, B., and Kemper, W. (1974). High activity of choline acetyltransferase induced in neuroblastoma × glia hybrid cells. *Exp. Cell Res.* **85,** 399–408.

Ames, B. N. (1979). Identifying environmental chemicals causing mutations and cancer. *Science* **204,** 587–593.

Anderson, R., Brown, M. S., and Goldstein, J. L. (1977). Role of the coated endocytic vesicle in the uptake of receptor-bound low density lipoprotein in human fibroblasts. *Cell* **10,** 351–364.

Arlett, C., and Harcourt, S. (1972). Expression time and spontaneous mutability in the estimation of induced mutation frequency following treatment of Chinese hamster cells by ultraviolet light. *Mutat. Res.* **16,** 301–306.

Arlett, C., and Potter, J. (1971). Mutation to 8-azaguanine resistance induced by gamma-radiation in a Chinese hamster cell line. *Mutat. Res.* **13,** 59–65.

Arlett, C., Turnbull, D., Harcourt, S., Lehmann, A., and Coletta, C. (1975). A comparison of the 8-azaguanine and ouabain-resistance systems for the selection of induced mutant Chinese hamster cells. *Mutat. Res.* **13,** 261–278.

Askanas, V., and Engel, W. (1975). A new program for investigating adult human skeletal muscle grown aneurally in tissue culture. *Neuroscience* **25,** 58–67.

Avery, O. T., MacLeod, C. M., and McCarty, M. (1944). Studies on the chemical nature of the substance inducing transformation of pneumococcal types. *J. Exp. Med.* **79,** 137–158.

225

Aviles, D., Ritz, E., and Jami, J. (1980). Chromosomes in tumors derived from mouse tumor × diploid cell hybrids obtained *in vitro. Somatic Cell Genet.* **6**, 171–186.

Azumi, J., and Sachs, L. (1977). Chromosome mapping of the genes that control differentiation and malignancy in myeloid leukemic cells. *Proc. Natl. Acad. Sci. U.S.A.* **74**, 253–257.

Badger, K., and Sussman, H. (1976). Structural evidence that human liver and placental alkaline phosphatase isozymes are coded by different genes. *Proc. Natl. Acad. Sci. U.S.A.* **73**, 2201–2205.

Baglioni, C. (1962). The fusion of two peptide chains in hemoglobin lepore and its interpretation as a genetic delection. *Proc. Natl. Acad. Sci. U.S.A.* **48**, 1880–1886.

Baird, M. B., Samis, H. V., Massie, H. R., and Zimmerman, J. A. (1975). A brief argument in opposition to the Orgel hypothesis. *Gerontologia* **21**, 57–63.

Baker, R., Burnett, D., Manowitz, R., Thompson, L., Siminovitch, L., and Till, J. (1974). Ouabain resistant mutants of mouse and hamster cells in culture. *Cell* **1**, 9–21.

Baker, W. K. (1968). Position-effect varigation. *Adv. Genet.* **14**, 133–169.

Barnstable, C. (1980). Monoclonal antibodies which recognize different cell types in the rat retina. *Nature (London)* **286**, 231–235.

Barnstable, C., Bodmer, W., Brown, G., Galfare, G., Milstein, C., Williams, A., and Zigler, A. (1978). Production of monoclonal antibodies to group A erythrocytes, HLA and other human cell surface antigens—new tools for genetic analysis. *Cell* **14**, 9–20.

Barrett, J. C., and Ts'o, P. O. P. (1978). Relationship between somatic mutation and neoplastic transformation. *Proc. Natl. Acad. Sci. U.S.A.* **75**, 3297–3301.

Barski, G., and Cornefert, F. (1962). Characteristics of ''hybrid''-type clonal cell lines obtained from mixed culture *in vitro. JNCI, J. Natl. Cancer Inst.* **28**, 801–821.

Barski, G., Sorieul, S., and Cornefert, F. (1960). Production dans des cultures *in vitro* de deux souches cellulaires en association de cellules de caractère ''hybride.'' *C.R. Hebd. Seances Acad. Sci.* **251**, 1825–1827.

Barski, G., Blanchard, M.-G., Youn, J. K., and Leon, B. (1973). Expression of malignancy in interspecies Chinese hamster × mouse cell hybrids. *JNCI, J. Natl. Cancer Inst.* **51**, 781–792.

Barton, R., Waters, L., and Yang, W. (1975). *In vitro* DNA synthesis by low molecular weight DNA polymerase. Increased infidelity associated with aging. *Fed. Proc., Fed. Am. Soc. Exp. Biol.* **33**, 1419 (Abstr. No. 1102).

Basilico, C. (1977). Temperature sensitive mutations in animal cells. *Adv. Cancer Res.* **24**, 223–266.

Baxter, J., and Funder, J. (1979). Hormone receptors. *N. Engl. J. Med.* **301**, 1149–1161.

Bazzell, K., Price, G., Tu, S., Griffin, M., Cox, R., and Ghosh, N. (1976). Cortisol modification of HeLa 65 alkaline phosphatase. Decreased phosphate content of the induced enzyme. *Eur. J. Biochem.* **61**, 493–499.

Beaumont, A., and Hughes, J. (1979). Biology of opioid peptides. *Annu. Rev. Pharmacol. Toxicol.* **19**, 245–267.

Bech-Hansen, N. T., Till, J. E., and Ling, V. (1976). Pleiotropic phenotype of colchicine-resistant CHO cells: Cross resistance and collateral sensitivity. *J. Cell Physiol.* **88**, 23–32.

Belote, J. M., and Lucchesi, J. C. (1980). Control of X chromosome transcription by the maleness gene in *Drosophila. Nature (London)* **285**, 573–575.

Benda, P., Lightbody, J., Sato, G., Levine, L., and Sweet, W. (1968). Differentiated rat glial cell strain in tissue culture. *Science* **161**, 370–371.

Benditt, E. P. (1977). The origin of atherosclerosis. *Sci. Am.* **236**(2), 74–85.

Benedict, W. F., Nebert, D. W., and Thompson, E. B. (1972). Expression of aryl hydroxylase induction and suppression of tyrosine aminotransferase induction in somatic-cell hybrids. *Proc. Natl. Acad. Sci. U.S.A.* **69**, 2179–2183.

Benoff, S., and Skoultchi, A. (1977). X-linked control of hemoglobin production in somatic hybrids of mouse erythroleukemic cells and mouse lymphoma or bone marrow cells. *Cell* **12**, 263–274.

Benoff, S., Bruce, S., and Skoultchi, A. (1978). Negative control of hemoglobin production in somatic cell hybrids due to heme deficiency. *Proc. Natl. Acad. Sci. U.S.A.* **75**, 4354–4358.

Bernard, O., Hozumic, N., and Tonegawa, S. (1978). Sequences of mouse immunoglobulin light chain genes before and after somatic changes. *Cell* **15**, 1133–1144.

Bernhard, H. P., Darlington, G. J., and Ruddle, F. H. (1973). Expression of liver phenotypes in cultured mouse hepatoma cells: Synthesis and secretion of serum albumin. *Dev. Biol.* **35**, 83–96.

Bernstein, I., Tam, M., and Nowinski, R. (1980). Mouse leukemia therapy with monoclonal antibodies against a thymus differentiation antigen. *Science* **207**, 68–71.

Bertolotti, R. (1979). Expression of differentiated functions in hepatoma cell hybrids: Selection in glucose free media of segregated hybrid cells which re-express gluconeogenic enzymes. *Somatic Cell Genet.* **3**, 579–602.

Bertolotti, R., and Weiss, M. C. (1972a). Expression of differentiated functions in hepatoma cell hybrids. II. Aldolase. *J. Cell. Physiol.* **79**, 221–233.

Bertolotti, R., and Weiss, M. C. (1972b). Expression of differentiated functions in hepatoma cell hybrids. VI. Extinctions and re-expression of liver alcohol dehydrogenase. *Biochimie* **54**, 195–210.

Bertolotti, R., and Weiss, M. C. (1972c). Aldolase in hepatoma cell hybrids: Extinction of the hepatic form and re-expression following loss of chromosomes. *In* "Cell Differentiation" (R. Harris, P. Allin, and D. Viza, eds.), pp. 202–205. Munksgaard, Copenhagen.

Bertolotti, R., and Weiss, M. C. (1974). Expression of differentiated functions in hepatoma cell hybrids. V. Re-expression of aldolase *in vitro* and *in vivo*. *Differentiation* **2**, 5–7.

Biedler, J. L., and Spengler, B. A. (1976). Metaphase chromosome anomaly: Association with drug resistance and cell specific products. *Science* **191**, 185–187.

Biquard, J. (1974). Agglutination by concanavalin A of normal chick embryo fibroblasts treated by 5-bromodeoxyuridine. *J. Cell. Physiol.* **84**, 459–462.

Bishop, J. (1981). Enemies within: The genesis of retrovirus oncogenes. *Cell* **23**, 5–6.

Bishop, J., and Jones, K. (1972). Chromosomal localization of human hemoglobin structural genes. *Nature* **240**, 149–150.

Bjorklund, A., and Stenevi, U. (1979). Reconstruction of the nigrostriatal dopamine pathway by intracerebral nigral transplants. *Brain Res.* **177**, 555–560.

Blau, H., and Epstein, C. (1979). Manipulation of myogenesis *in vitro:* Reversible inhibition by DMSO. *Cell* **17**, 95–108.

Boersma, D., McGill, S., Mollenkamp, J., and Roufa, D. J. (1979). Emetine resistance in Chinese hamster cells: Analysis of ribosomal proteins prepared from mutant cells. *J. Biol. Chem.* **254**, 559–567.

Bollon, A. (1980). Analysis of yeast *ilv-1* cis control and domain mutants. *Mol. Gen. Genet.* **177**, 283–289.

Bolund, L., Ringertz, N., and Harris, H. (1969). Changes in the cytochemical properties of erythrocyte nuclei reactivated by cell fusion. *J. Cell Sci.* **4**, 71–87.

Born, G., Grutzner, P., and Hemminger, H. (1976). Evidenz für eine mosaikstruktur der netzhaut bei konduktorinnen für dichromasie. *Hum. Genet.* **32**, 189–196.

Bourne, H., Coffino, P., and Tomkins, G. (1975). Somatic genetic analysis of cyclic AMP action: Selection of unresponsive mutants. *J. Cell. Physiol.* **85**, 603–610.

Boveri, T. (1920). "The Origin of Malignant Tumors." Williams & Wilkins, Baltimore, Maryland.

Bradley, W. E. (1979). Reversible inactivation of autosomal alleles in Chinese hamster cells. *J. Cell Physiol.* **101**, 325–340.

Bradley, W. E., and Letovanec, D. (1982). High frequency non-random mutational event at the adenine phosphoribosyltransferase (APRT) locus of sib-selected CHO variants for APRT. *Somatic Cell Genet.* **8**, 51–66.

Breen, G., and Scheffler, I. (1980). Cytoplasmic inheritance of oligomycin resistance in Chinese hamster ovary cells. *J. Cell Biol.* **86**, 723–729.

Brenner, M., Tisdale, P., and Loomis, W. F., Jr. (1975). Techniques for rapid biochemical screening of large numbers of cell clones. *Exp. Cell Res.* **90**, 249–252.

Breslow, R. E., and Goldsby, R. A. (1969). Isolation and characterization of thymidine transport mutants of Chinese hamster cells. *Exp. Cell Res.* **55**, 339–346.

Bridges, B., and Huckle, J. (1970). Mutagenesis of cultured mammalian cells by X-radiation and ultraviolet light. *Mutat. Res.* **10**, 141–151.

Briggs, R., and King, T. (1952). Transplantation of living nuclei from blastula cells into enucleated frogs eggs. *Proc. Natl. Acad. Sci. U.S.A.* **38**, 455–463.

Brown, J. E., and Weiss, M. C. (1975). Activation of production of mouse liver enzymes in rat hepatoma–mouse lymphoid cell hybrids. *Cell* **6**, 481–494.

Brumbaugh, J., and Lee, K. (1975). The gene action and function of two DOPA oxidase positive melanocyte mutants of the fowl. *Genetics* **81**, 333–347.

Buck, C., and Bodmer, W. (1975). Rotterdam conference (1974) second international workshop on human gene mapping. *Birth Defects, Orig. Artic. Ser.,* **11**(3), 87–89.

Bulmer, D., Stocco, D., and Morrow, J. (1976). Bromodeoxyuridine induced variations in the level of alkaline phosphatase in several human heteroploid cell lines. *J. Cell. Physiol.* **87**, 357–366.

Bunn, C., and Tarrant, G. (1980). Limited life span in somatic cell hybrids and cybrids. *Exp. Cell Res.* **127**, 385–396.

Burch, J. W., and McBride, O. W. (1975). Human gene expression in rodent cells after uptake of isolated metaphase chromosomes. *Proc. Natl. Acad. Sci. U.S.A.* **72**, 1797–1801.

Burch, P. (1979). Coronary disease: Risk factors, age and time. *Am. Heart J.* **97**, 415–419.

Burch, P. R., and Jackson, D. (1976). Molecular mechanisms of ageing. A critique. *Gerontology* **22**, 206–211.

Burdon, R. H., and Adams, R. L. P. (1980). Eukaryotic DNA methylation. *Trends Biochem. Sci.* **5**, 294–297.

Burnett, M. (1959). "The Clonal Selection Theory of Acquired Immunity." Vanderbilt Univ. Press, Nashville, Tennessee.

Burns, R. L., Rosenberger, P. G., and Klebe, R. J. (1976). Carbohydrate preferences of mammalian cells. *J. Cell. Physiol.* **88**, 307–316.

Burton, E., and Metzenburg, R. (1972). Novel mutation causing derepression of several enzymes of sulfur metabolism in *Neurospora crassa. J. Bacteriol.* **109**, 140–151.

Busch, D., Cleaver, J., and Glaser, D. (1980). Large scale isolation of UV-sensitive clones of CHO cells. *Somatic Cell Genet.* **6**, 407–418.

Butenko, R. (1979). Cultivation of isolated protoplasts and hybridization of somatic plant cells. *Int. Rev. Cytol.* **59**, 323–366.

Caboche, M. (1974). Comparison of the frequencies of spontaneous and chemically induced 5-bromodeoxyuridine-resistant mutations in wild-type and revertant BHK-21/13 cells. *Genetics* **77**, 309–322.

Cairns, J. (1978). "Cancer, Science and Society." Freeman, San Francisco, California.

Cairns, J. (1981). The origin of human cancers. *Nature (London)* **289**, 353–357.

Capecchi, M. R. (1980). High efficiency transformation by direct microinjection of DNA into cultured mammalian cells. *Cell* **22**, 479–488.

Capecchi, M. R., Vonderhaar, R. A., Capecchi, N. E., and Sveda, M. M. (1977). The isolation of a suppressible nonsense mutant in mammalian cells. *Cell* **12**, 371–381.

Capizzi, R. L., and Jameson, J. J. (1973). A table for the estimation of the spontaneous mutation rate of cells in culture. *Mutat. Res.* **17**, 147–148.

Carlson, P. (1973). The use of protoplasts for genetic research. *Proc. Natl. Acad. Sci. U.S.A.* **70,** 598–602.

Carlsson, S., Luger, O., Ringertz, N., and Savage, R. (1974a). Phenotypic expression in chick erythrocyte × rat myoblast hybrids and in chick myoblast × rat myoblast hybrids. *Exp. Cell Res.* **84,** 47–55.

Carlsson, S., Ringertz, N., and Savage, R. (1974b). Intracellular antigen migration in interspecific myoblast heterokaryons. *Exp. Cell Res.* **84,** 255–266.

Carritt, B., Goldfarb, D. S. G., Hooper, M. L., and Slack, C. (1977). Chromosome assignment of a human gene for argininosuccinate synthetase expression in Chinese hamster × human somatic cell hybrids. *Exp. Cell Res.* **106,** 71–78.

Carson, M. P., Vernick, D., and Morrow, J. (1974). Clones of Chinese hamster cells cultivated *in vitro* not permanently resistant to azaguanine. *Mutat. Res.* **24,** 47–54.

Caskey, C. T., and Kruh, G. D. (1979). The HPRT locus. *Cell* **16,** 1–9.

Cavalli-Sforza, L. L., and Bodmer, W. F. (1971). ''The Genetics of Human Populations.'' Freeman, San Francisco, California.

Chalazonitis, A., Greene, L. A., and Shain, W. (1975). Excitability and chemosensitivity properties of a somatic cell hybrid between mouse neuroblastoma and sympathetic ganglion cells. *Exp. Cell Res.* **96,** 225–238.

Chan, V. L., Whitmore, G. F., and Siminovitch, L. (1972). Mammalian cells with altered forms of RNA polymerase II. *Proc. Natl. Acad. Sci. U.S.A.* **69,** 3119–3123.

Chang, C., Phillips, C., Trosko, J., and Hart, R. (1977). Mutagenic and epigenetic influence of caffeine on the frequencies of UV-induced ouabain-resistant Chinese hamster cells. *Mutat. Res.* **45,** 125–136.

Chasin, L. A. (1973). The effect of ploidy on chemical mutagenesis in cultured Chinese hamster cells. *J. Cell. Physiol.* **82,** 299–308.

Chasin, L. A. (1974). Mutations affecting adenine phosphoribosyl transferase activity in Chinese hamster cells. *Cell* **2,** 37–41.

Chasin, L. A., and Urlaub, G. (1976). Mutant alleles for hypoxanthine phosphoribosyltransferase: Codominant expression, complementation and segregation in hybrid Chinese hamster cells. *Somatic Cell Genet.* **2,** 453–467.

Chen, L. B. (1977). Alterations in cell surface LETS protein during myogenesis. *Cell* **10,** 393–400.

Choo, K. H., and Cotton, R. G. H. (1977). Genetics of the mammalian phenylalanine hydroxylase system. I. Isolation of phenylalanine hydroxylase deficient tyrosine auxotrophs from rat hepatoma cells. *Somatic Cell Genet.* **3,** 457–470.

Choo, K. H., and Cotton, R. G. H. (1979). Genetics of the mammalian phenylalanine hydroxylase system. II. Immunological and two-dimensional gel electrophoretic studies of phenylalanine hydroxylase in cultured normal and mutant rat hepatoma cells. *Biochem. Genet.* **17,** 921–946.

Chorazy, M., Bendich, A., Borenfreund, E., Ittensohn, O., and Hutchison, I. (1963). Uptake of mammalian chromosomes by mammalian cells. *J. Cell Biol.* **19,** 71–77.

Chu, E. H., Sun, N. C., and Chang, C. C. (1972). Induction of auxotrophic mutations by treatment of Chinese hamster cells with 5-bromodeoxyuridine and black light. *Proc. Acad. Sci. U.S.A.* **69,** 3459–3463.

Chu, E. H. Y. (1971). Mammalian cell genetics. III. Characterization of x-ray induced forward mutations in Chinese hamster cell cultures. *Mutat. Res.* **11,** 23–24.

Chu, E. H. Y., and Malling, H. (1968). Mammalian cell genetics. II. Chemical induction of specific locus mutations in Chinese hamster cells *in vitro*. *Proc. Natl. Acad. Sci. U.S.A.* **61,** 1306–1312.

Chu, E. H. Y., Brimer, P., Jacobson, K. B., and Merriam, E. V. (1969). Mammalian cell genetics. I. Selection and characterization of mutations auxotrophic for L-glutamine or resistant to 8-azaguanine in Chinese hamster cells *in vitro*. *Genetics* **62,** 359–377.

Cifone, M. A., Hynes, R. O., and Baker, R. M. (1979). Characteristics of concanavalin A resistant Chinese hamster ovary cells and certain revertants. *J. Cell. Physiol.* **100,** 39–54.

Claverie, J. M., De Souza, A. C., and Thirion, J. P. (1979). Mutations of Chinese hamster somatic cells from 2-deoxygalactose sensitivity to resistance. *Genetics* **92,** 563–572.

Cleaver, J. E., and Bootsma, D. (1975). Xeroderma pigmentosum: biochemical and genetic characteristics. *Annu. Rev. Genet.* **9,** 19–38.

Cline, M. H., Stang, H., Mercola, K., Morse, L., Ruprecht, R., Brown, J., and Salzer, W. (1980). Gene transfer in intact animals. *Nature (London)* **284,** 422–425.

Clive, D., and Moore-Brown, M. M. (1979). The L5178Y/TK $^{+/-}$ mutagen assay system: Mutant analysis. *In* "Banbury Report 2: Mammalian Cell Mutagenesis: The Maturation of Test Systems" (A. Hsie, J. P. O'Neill, and V. McElheny, eds.), pp. 421–430. Cold Spring Harbor Lab., Cold Spring Harbor, New York.

Clive, D., and Voytek, P. (1977). Evidence for chemically-induced structural gene mutations at the thymidine kinase locus in cultured L5178Y mouse lymphoma cells. *Mutat. Res.* **44,** 269–278.

Clive, D., Flamm, W. G., Machesko, M. R., and Bernheim, N. J. (1972). A mutational assay system using the thymidine kinase locus in mouse lymphoma cells. *Mutat. Res.* **16,** 77–87.

Cocking, E., Davey, M., Pental, D., and Power, J. (1981). Aspects of plant genetic manipulation. *Nature (London)* **293,** 265–269.

Codish, S. D., and Paul, B. (1974). Reversible appearance of a specific chromosome which suppresses malignancy. *Nature (London)* **252,** 610–612.

Coffino, P., and Scharff, M. (1971). Rate of somatic mutation in immunoglobulin production by mouse myeloma cells. *Proc. Natl. Acad. Sci. U.S.A.* **68,** 219–233.

Coffino, P., Knowles, B., Nathenson, S., and Scharff, M. (1971). Suppression of immunoglobulin synthesis by cellular hybridization. *Nature (London)* **231,** 87–90.

Coffino, P., Bourne, H. R., and Tompkins, G. M. (1975). Somatic genetic analysis of cyclic AMP action: Selection of unresponsive mutants. *J. Cell. Physiol.* **85,** 603–610.

Coleclough, C., Perry, R., Karjalainen, K., and Weigert, M. (1981). Aberrant rearrangements contribute significantly to the allelic exclusion of immunoglobin gene expression. *Nature (London)* **290,** 372–378.

Coleman, J., and Coleman, A. (1968). Muscle differentiation and macromolecular synthesis. *J. Cell. Physiol.* **72,** 19–34.

Collier, J. R. (1975). Diphtheria toxin: Mode of action and structure. *Bacteriol. Rev.* **39,** 54–85.

Colofiore, J., Morrow, J., and Patterson, M. K., Jr. (1973). Asparagine requiring tumor cell lines and their non-requiring variants: Cytogenetics, biochemistry and population dynamics. *Genetics* **75,** 503–514.

Comfort, A. (1979). "The Biology of Senescence." Elsevier/North-Holland, New York.

Comings, D. E. (1973). A general theory of carcinogenesis. *Proc. Natl. Acad. Sci. U.S.A.* **70,** 3324–3328.

Cory, S., and Adams, J. (1980). Deletions are associated with somatic rearrangement of immunoglobulin heavy chain genes. *Cell* **19,** 37–51.

Cotton, R., and Milstein, C. (1973). Fusion of two immunoglobulin-producing myeloma cells. *Nature (London)* **244,** 42–43.

Cox, F. (1980). Monoclonal antibodies and immunity to malaria. *Nature (London)* **284,** 304–304.

Cox, R., and Masson, W. (1978). Do radiation-induced thioguanine-resistant mutants of cultured mammalian cells arise by HGPRT gene mutation or X-chromosome rearrangement? *Nature (London)* **276,** 629–630.

Cox, R., Krauss, M. R., Balis, M. E., and Dancis, J. (1972). Communication between normal and enzyme deficient cells in tissue culture. *Exp. Cell Res.* **74,** 251–268.

Cox, R. B., and King, J. C. (1975). Gene expression in cultured mammalian cells. *Int. Rev. Cytol.* **43,** 282–353.

Cox, R. P., and MacLeod, C. M. (1962). Alkaline phosphatase content and the effects of pred-nisolone on mammalian cells in culture. *J. Gen. Physiol.* **45**(1), 439–485.

Croce, C. M. (1977). Assignment of the integration site for simian virus 40 to chromsome 17 in GM54VA, a human cell line transformed by simian virus 40. *Proc. Natl. Acad. Sci. U.S.A.* **74**, 315–318.

Croce, C. M., Litwalk, G., and Koprowski, H. (1973a). Human regulatory gene for inducible tyrosine aminotransferase in rat–human hybrids. *Proc. Natl. Acad. Sci. U.S.A.* **70**, 1268–1272.

Croce, C. M., Bakay, B., Nijhan, W. L., and Koprowski, H. (1973b). Re-expression of the rat hypoxanthinephosphoribosyl transferase gene in rat–human hybrids. *Proc. Natl. Acad. Sci. U.S.A.* **70**, 2590–2594.

Croce, C. M., Koprowski, H., and Litwalk, G. (1974a). Regulation of the corticosteroid inducibility of tyrosine aminotransferase in interspecific hybrid cell. *Nature (London)* **249**, 839–841.

Croce, C. M., Litwalk G., and Koprowski, H. (1974b). Regulation of the inducibility of tyrosine aminotransferase in rat–human hybrids. *In* "Somatic Cell Hybridization" (R. L. Davidson and F. F. de la Cruz, eds.), pp. 173–176. Raven, New York.

Croce, C., Shander, M., Martinis, J., Cicurel, L., D'Ancona G., Dolby, T., and Koprowski, H. (1979a). Chromosomal location of the genes for human immunoglobulin heavy chains. *Proc. Natl. Acad. Sci. U.S.A.* **76**, 3416–3419.

Croce, C. M., Barrick, J., Linnenbach, A., and Koprowski, H. (1979b). Expression of malignancy in hybrids between normal and malignant cells. *J. Cell. Physiol.* **99**, 279–286.

Danes, B. S., and Bearn, A. G. (1967). Hurler's syndrome: A genetic study of clones in cell culture with particular reference to the Lyon hypotheses. *J. Exp. Med.* **126**, 509–522.

Daniel, C. W., DeOme, K. B., Young, J., Blair, P., and Faulkin, L. (1968). The *in vivo* life span of normal and preoplastic mouse mammary glands: A serial transplantation study. *Proc. Natl. Acad. Sci. U.S.A.* **61**, 53–60.

Daniels, M., and Hamprecht, B. (1974). The ultrastructure of neuroblastoma glioma somatic cell hybrids. *J. Cell Biol.* **63**, 691–699.

Darlington, G. J., Bernhard, H. P., and Ruddle, F. H. (1974a). The expression of hepatic functions in mouse hepatoma × human leukocyte hybrids. *Cytogenetics* **13**, 86–88.

Darlington, G. J., Bernhard, H. P., and Ruddle, F. H. (1974b). Human serum albumin phenotype activation in mouse hepatoma–human leukocyte cell hybrids. *Science* **185**, 859–862.

Darlington, G. J., Bernhard, H. P., and Ruddle, F. H. (1975). Expression of hepatic functions in somatic cell hybrids. *In* "Gene Expression and Carcinogenesis in Cultured Liver Cells" (L. E. Gershenson and E. B. Thompson, eds.), pp. 333–345. Academic Press, New York.

Davidson, E., Hough, B., Klem, W., and Britten, R. (1975). Structural genes adjacent to in-terspersed repetitive DNA sequences. *Cell* **4**, 217–238.

Davidson, J. N., and Patterson, D. (1979). Alteration in structure of multifunctional protein from Chinese hamster ovary cells defective in pyrimidine biosynthesis. *Proc. Natl. Acad. Sci. U.S.A.* **76**, 1731–1735.

Davidson, R. (1973). Somatic cell hybridization: Studies on genetics and development. *In* "Addison-Wesley Modules in Biology." Addison-Wesley, Reading, Massachusetts.

Davidson, R. (1974). Control of expression of differentiated functions in somatic cell hybrids. *In* "Somatic Cell Hybridization" (R. Davidson and F. F. de la Cruz, eds.), pp. 131–150. Raven, New York.

Davidson, R., and Ephrussi, B. (1965). A selective system for the isolation of hybrids between L cells and normal cells. *Nature (London)* **205**, 1170–1171.

Davidson, R., and Kaufman, E. (1977). Deoxycytidine reverses the suppression of pigmentation caused by 5'-BrdUrd without changing the amount of 5-BrdUrd in DNA. *Cell* **12**, 923–929.

Davidson, R. G., Nitowsky, H. M., and Childs, B. (1963). Demonstration of two populations of cells in the human female heterozygous for glucose-6-phosphate dehydrogenase variants. *Proc. Natl. Acad. Sci. U.S.A.* **50**, 481–485.

Davidson, R. L., and Gerald, P. S. (1976). Improved technique for the induction of mamalian cell hybridization by polyethylene glycol. *Somatic Cell Genet.* **2**, 165–176.

Davies, K. E., Young, B. D., Elles, R. G., Hill, M. E., and Williamson, R. (1981). Cloning of a representative genomic library of the human X chromosome after sorting by flow cytometry. *Nature (London)* **293**, 374–376.

Davis, F. M., and Adelberg, E. A. (1973). Use of somatic cell hybrids for analysis of the differentiated state. *Bacteriol. Rev.* **37**, 197–214.

Davis, M., Calame, K., Early, P., Livant, D., Joho, R., Weissman, I., and Hood, L. (1980). An immunoglobulin heavy-chain gene is formed by at least two recombinational events. *Nature (London)* **283**, 733–739.

Degnen, G. E., Miller, J. L., Eisenstadt, J. M., and Adelberg, E. A. (1976). Chromosome-mediated gene transfer between closely related strains of cultured mouse cells. *Proc. Natl. Acad. Sci. U.S.A.* **73**, 2838–2842.

Deisseroth, A., and Hendrick, D. (1979). Activation of phenotypic expression of human globin genes from nonerythroid cells by chromosome-dependent transfer to tetraploid mouse erythroleukemia cells. *Proc. Natl. Acad. Sci. U.S.A.* **76**, 2185–2189.

Deisseroth, A., Velez, R., and Nienhuis, A. (1976). Hemoglobin synthesis in somatic cell hybrids: Independent segregation of the human alpha- and beta-globin genes. *Science* **191**, 1262–1264.

Deisseroth, A., Velez, R., Burk, R., Minna, J., Anderson, W. F., and Nienhuis, A. (1978). Extinction of globin gene expression of human fibroblast × mouse erythroleukemia cell hybrids. *Somatic Cell Genet.* **2**, 373–384.

Dell'Orco, R., Mertens, J., and Kruse, P., Jr. (1973). Doubling potential calendar time and senescence of human diploid cells in culture. *Exp. Cell Res.* **77**, 356–360.

DeMars, R. (1974). Resistance of cultured human fibroblasts and other cells to purine and pyrimidine analogues in relation to mutagenesis detection. *Mutat. Res.* **24**, 335–364.

DeMars, R., and Held, K. (1972). The spontaneous azaguanine resistant mutants of diploid human fibroblasts. *Humangenetik* **16**, 87–110.

DeMars, R., and Hooper, J. (1960). A method of selecting for auxotrophic mutants of HeLa cells. *J. Exp. Med.* **111**, 559–573.

de Saint Vincent, B. R., and Buttin, G. (1979). Studies on 1-beta-D-arabinofuranosyl cytosine-resistant mutants of Chinese hamster fibroblasts. III. Joint resistance to arabinofurnanosyl cytosine and to excess thymidine a semidominant manifestation of deoxycytidine triphosphate pool expansion. *Somatic Cell Genet.* **5**, 67–82.

D'Eustachio, P., Pravtcheva, D., Marcu, K., and Ruddle, F. (1980). Chromosomal location of the structural gene cluster encoding murine immunoglobulin heavy chains. *J. Exp. Med.* **151**, 1545–1550.

DeVellis, J., and Inglish, D. (1969). Effects of ionizing radiation on hormone induction of enzymes in developing and adult rat brain and in an established line of glial cells. *Radiat. Res.* **39**, 488–489.

Devoret, R. (1979). Bacterial tests for potential carcinogens. *Sci. Am.* **241**, 40–49.

Diacumakos, E. (1973). Methods for micromanipulation of human somatic cells in culture. *Methods Cell Biol.* **7**, 287–311.

Dickerman, L. H., and Tischfield, J. (1978). Comparative effects of adenine analogs upon metabolic cooperation between Chinese hamster cells with different levels of adenine phosphoribosyl transferase activity. *Mutat. Res.* **49**, 83–94.

Doerson, C., and Stanbridge, E. (1979). Cytoplasmic inheritance of erythromycin resistance in human cells. *Proc. Natl. Acad. Sci. U.S.A.* **76**, 4549–4553.

Dreyer, W., and Bennett, J. (1965). The molecular basis of antibody formation: A paradox. *Proc. Natl. Acad. Sci. U.S.A.* **54**, 864–869.

Dufresne, M., Rogers, J., Coulter, M., Ball, E., Lo, T., and Sanwal, B. D. (1976). Apparent dominance of serine auxotrophy and the absence of expression of muscle-specific proteins in rat myoblast × mouse L-cell hybrids. *Somatic Cell Genet.* **2,** 521–535.

Duncan, M., Dell'Orco, R., and Guthrie, P. (1972). Relationship of heat labile glucose-6-phosphate dehydrogenase and multiple forms of the enzyme in senescent human fibroblasts. *J. Cell. Physiol.* **93,** 49–56.

Dym, H., Turner, D. C., Eppenberger, H. M., and Yaffe, D. (1978). Creatine kinase isoenzyme transition in actinomycin D-treated differentiating muscle cultures. *Exp. Cell Res.* **113,** 15–21.

Eagle, H. (1959). Amino acid metabolism in mammalian cell cultures. *Science* **130,** 432–437.

Ellis, G., and Goldberg, D. M. (1972). Assay of human serum and liver guanase activity with 8-azaguanine as a substrate. *Clin. Chim. Acta* **37,** 47–52.

Elson, N. A., and Cox, R. P. (1969). Production of fetal-like alkaline phosphatase by HeLa cells. *Biochem. Genet.* **3,** 549–561.

Endo, A., Kuroda, M., and Tanzawa, K. (1976). Competitive inhibition of 3-hydroxy-3-methyl-glutaryl coenzyme A reductase by ML236A and ML236B fungal antimetabolites, having hypocholesterolemic activity. *FEBS Lett.* **72,** 323–326.

Ephrussi, B., and Weiss, M. (1969). Hybrid somatic cells. *Sci. Am.* **220**(4), 26–35.

Ephrussi, B., Davidson, L. R., and Weiss, M. C. (1969). Malignancy of somatic cell hybrids. *Nature (London)* **224,** 1314–1315.

Epstein, J., Williams, J., and Little, J. (1974). Rate of DNA repair in progeric and normal human fibroblasts. *Biochem. Biophys. Res. Commun.* **59,** 850–857.

Esko, J. D., and Raetz, C. R. H. (1978). Replica plating and *in situ* enzymatic assay of animal cell colonies established on filter paper. *Proc. Natl. Acad. Sci. U.S.A.* **75,** 1190–1193.

Evain, D., Gottesman, M., Pastan, I., and Anderson, W. (1979). A mutation affecting the catalytic subunit of cyclic AMP-dependent protein kinase in CHO cells. *J. Biol. Chem.* **254,** 6931–6937.

Farrell, S. A., and Worton, C. G. (1977). Chromosome loss is responsible for segregation at the HPRT locus in Chinese hamster cell hybrids. *Somatic Cell Genet.* **3,** 539–551.

Fasy, T., Cullen, B., Luk, D., and Bick, M. (1980). Studies on the enhanced interaction of halo-deoxyuridine-substituted DNAs with H1 histones and other polypeptides. *J. Biol. Chem.* **255,** 1380–1387.

Fazekas de St. Groth, S., and Scheidiegger, D. (1980). Production of monoclonal antibodies: strategy and tactics. *J. Immunol. Methods* **35,** 1–21.

Fellous, M., Bengtsson, B., Finnegan, D., and Bodmer, W. F. (1974). Expression of the Xg[a] antigen on cells in culture, and its segregation in somatic cell hybrids. *Ann. Hum. Genet.* **37,** 421–430.

Fenwick, R. G., Jr., Sawyer, T. H., Kruh, G. D., Astrin, K. H., and Caskey, C. T. (1977). Forward and reverse mutations affecting the kinetics and apparent molecular weight of mammalian HGPRT. *Cell* **12,** 383–391.

Fergusen-Smith, M. A. (1966). X–Y chromosomal interchange in the aetiology of Klinefelter's syndrome. *Lancet* **2,** 475–476.

Fialkow, P. J., Gartler, S. M., and Yoshida, A. (1967). Clonal origin of chronic myelocytic leukemia in man. *Proc. Natl. Acad. Sci. U.S.A.* **58,** 1468–1471.

Fluck, R. A., and Strohman, R. C. (1973). Acetylcholinesterase activity in developing skeletal muscle cells *in vitro. Dev. Biol.* **33,** 417–428.

Fougere, C., and Weiss, M. (1978). Phenotypic exclusion in mouse melanoma–rat hepatoma hybrid cells: Pigment and albumin production are not reexpressed simultaneously. *Cell* **15,** 843–854.

Fournier, R. E. K., and Ruddle, F. H. (1977). Microcell-mediated chromosome transfer. *ICN–UCLA Symp. Mol. Cell. Biol.* **7,** 189–199.

Fox, M. (1974). The effect of post treatment with caffeine on survival and uv-induced mutation frequencies in Chinese hamster and mouse lymphoma cells *in vitro. Mutat. Res.* **24,** 187–204.

Fox, M., and Radacic, M. (1978). Adaptational origin of some purine-analogue resistant phenotypes in cultured mammalian cells. *Mutat. Res.* **49,** 275–296.

Franceschetti, A., and Klein, D. (1954). Le depistage des heterozygotes. *In* "Genetica medica" (L. Gedda, ed.). Orrizonte Medico, Rome.

Franks, L. (1970). Cellular aspects of aging. *Exp. Gerontol.* **5,** 281–289.

Freed, J. J., and Hames, I. M. (1976). Loss of thermolabile thymidine kinase activity in bromodeoxyuridine resistant (transport deficient, kinase positive) haploid cultured frog cells. *Exp. Cell Res.* **99,** 126–134.

Freed, J. J., and Mezger-Freed, L. (1973). Origin of thymidine kinase deficient (TK⁻) haploid frog cells via an intermediate thymidine transport deficient (TT⁻) phenotype. *J. Cell. Physiol.* **82,** 199–212.

Friedrich, V., and Coffino, P. (1977). Mutagenesis in S49 mouse lymphoma cells: Induction of resistance to ouabain, 6-thioguanine, and dibutyryl cyclic AMP. *Proc. Natl. Acad. Sci. U.S.A.* **74,** 679–683.

Friend, C., Scher, W., Holland, J., and Sato, T. (1971). Hemoglobin synthesis in murine virus-induced leukemic cells *in vitro:* Stimulation of erythroid differentiation by DMSO. *Proc. Natl. Acad. Sci. U.S.A.* **68,** 378–382.

Frye, L. D., and Edidin, M. (1970). The rapid intermixing of cell surface antigens after formation of mouse human heterokaryons. *J. Cell Sci.* **7,** 319–335.

Fulder, S. J., and Holliday, R. (1975). A rapid rise in cell variants during the senescence of populations of human fibroblasts. *Cell* **6,** 67–73.

Gabridge, M., and Legator, M. (1969). A host mediated microbial assay for the detection of mutagenic compounds. *Proc. Soc. Exp. Biol. Med.* **130,** 831–834.

Gallant, J., and Palmer, L. (1979). Error propagation in viable cells. *Mech. Ageing Dev.* **10,** 27–38.

Gally, J. (1973). Structure of immunoglobulins. *In* "The Antigens" (M. Sela, ed.), Vol. 1, pp. 162–285. Academic Press, New York.

German, J. (1972). Genes which increase chromosomal instability in somatic cells and predispose to cancer. *Prog. Med. Genet.* **8,** 61–102.

Gershon, D., and Gershon, H. (1976). An evaluation of the error catastrophe theory of aging in light of recent experimental results. *Gerontology* **22,** 212–219.

Gilliland, D., Steplewski, Z., Collier, R., Mitchell, K., Chang, T., and Koprowski, H. (1980). Antibody directed cytotoxic agents. Use of monoclonal antibody to direct the action of toxin-A chains in colorectal carcinoma cells. *Proc. Natl. Acad. Sci. U.S.A.* **77,** 4539–4543.

Goldfarb, P. S., Carritt, B., and Hooper, M. L. (1977). The isolation and characterization of asparagine-requiring mutants of Chinese hamster cells. *Exp. Cell Res.* **104,** 357–367.

Goldfarb, P. S. G., Slalk, C., Subak-Sharpe, J. H., and Wright, E. D. (1974). Metabolic co-operation between cells in tissue culture. *Symp. Soc. Exp. Biol.* **28,** 463–484.

Goldsby, R. A., and Zipser, E. (1969). The isolation and replica plating of mammalian cell clones. *Exp. Cell Res.* **54,** 271–275.

Goldstein, J. L., and Brown, M. S. (1979). The LDL receptor locus and the genetics of familial hypercholesterolemia. *Annu. Rev. Genet.* **13,** 259–290.

Goldstein, J. L., Schrott, H. G., Hazard, W., Bierman, E., and Motulsky, A. G. (1973). Hyper-lipidemia in coronary heart disease. II. Genetic analysis of lipid levels in 176 families and delineation of a new inherited disorder, combined hyperlipidemia. *J. Clin. Invest.* **52,** 1544–1568.

Goodman, J., and Wang, A. (1978). Immunoglobulins: Structure and diversity. *In* "Basic and Clinical Immunology" (H. H. Fudenberg, D. P. Stites, J. L. Caldwell, and J. V. Wells, eds.), p. 36. Lange Med. Publ., Los Altos, California.

Gopalakrishnan, T. V., and Anderson, W. F. (1979). Epigenetic activation of phenylalanine hydroxylase in mouse erythroleukemia cells by the cytoplast of rat hepatoma cells. *Proc. Natl. Acad. Sci. U.S.A.* **76,** 3932–3936.

Gordon, J., and Ruddle, F. H. (1981). Mammalian gonadal determination and gametogenesis. *Science* **211**, 1265–1271.

Gottesman, M. (1980). Genetic approaches to cyclic AMP effects in cultured mammalian cells. *Cell* **22**, 329–330.

Gough, N., Kemp, D., Tyler, B., Adams, J., and Cory, S. (1980). Intervening squences divide the gene for the constant region of mouse immunoglobulin mu chains into segments, each encoding a domain. *Proc. Natl. Acad. Sci. U.S.A.* **77**, 554–558.

Graf, L. H., McRoberts, J. A., Harrison, T. M., and Martin, D. W., Jr. (1976). Increased PRPP synthetase activity in cultured rat hepatoma cells containing mutations in the hypoxanthine guanine phosphoribosyl transferase gene. *J. Cell. Physiol.* **88**, 331–342.

Granner, D., Diesterhart, M., Noguchi, T., Olson, P., Hargrove, J., and Volentine, G. (1979). Regulation of liver and HTC cell tyrosine aminotransferase mRNA by glucocorticoids and dibutylryl cyclic AMP. *In* "Hormones and Cell Culture" (G. Sato and R. Rose, eds.), pp. 889–904. Cold Spring Harbor Lab., Cold Spring Harbor, New York.

Graves, J. (1972). DNA synthesis in heterokaryons formed by fusion of mammalian cells from different species. *Exp. Cell Res.* **72**, 393–403.

Griffin, J., and Wilson, J. (1980). The syndromes of androgen resistance. *N. Engl. J. Med.* **302**, 198–209.

Griffin, M., and Ber, R. (1969). Cell cycle events in the hydrocortisone regulation of alkaline phosphatase in HeLa S_3 cells. *J. Cell Biol.* **40**, 297–304.

Griffin, M., and Cox, R. (1966). Studies on the mechanism of hormonal induction of alkaline phosphatase in human cell cultures. I. Effects of puromycin and actinomycin D. *J. Cell Biol.* **29**, 1–9.

Grouse, L., Schrier, B., Letendre, C., Zubairi, M., and Nelson, P. (1980). Neuroblastoma differentiation involves both the disappearance of old and the appearance of new poly(A)$^+$ messenger RNA sequences in polyribosomes. *J. Biol. Chem.* **255**, 3871–3877.

Gupta, K. C., and Taylor, M. W. (1978). A amanitin-resistant mutants of Chinese hamster ovary (CHO) cells. *Mutat. Res.* **49**, 95–101.

Gupta, R. S. (1980). Senescence of cultured human diploid fibroblasts. Are mutations responsible? *J. Cell Physiol.* **103**, 209–216.

Gupta, R. S., and Siminovitch, L. (1976). The isolation and preliminary characterization of somatic cell mutants resistant to the protein synthesis inhibitor emetine. *Cell* **9**, 213–219.

Gupta, R. S., and Siminovitch, L. (1978a). Isolation and characterization of mutants of human diploid fibroblasts resistant to diphtheria toxin. *Proc. Natl. Acad. Sci. U.S.A.* **75**, 3337–3340.

Gupta, R. S., and Siminovitch, L. (1978b). Mutants of CHO cells resistant to the protein synthesis inhibitor emetine: Genetic and biochemical characterization of second step mutants. *Somatic Cell Genet.* **4**, 77–93.

Gupta, R. S., and Siminovitch, L. (1980). Diptheria toxin resistance in Chinese hamster cells. Genetic and biochemical characteristics of the mutants affected in protein synthesis. *Somatic Cell Genet.* **6**(3), 361–379.

Gupta, R. S., Chan, D. Y. H., and Siminovitch, L. (1978a). Evidence for variation in the number of functional gene copies at the AMA^R locus in Chinese hamster cells. *J. Cell. Physiol.* **97**, 461–468.

Gupta, R. S., Chan, D. Y. H., and Siminovitch, L. (1978b). Evidence for functional hemizygosity at the EMt^r locus in CHO cells through segregation analysis. *Cell* **14**, 1007–1013.

Gurdon, J. (1967). On the origin and persistence of a cytoplasmic state inducing nuclear DNA synthesis in frog eggs. *Proc. Natl. Acad. Sci. U.S.A.* **58**, 545–552.

Gurdon, J. B. (1974). "The Control of Gene Expression in Animal Development." Harvard Univ. Press, Cambridge, Massachusetts.

Gusella, J., Varsanyi-Breiner, A., Kao, F., Jones, C., Puck, T., Keys, C., Orkin, S., and Housman, D. (1979). Precise localization of human beta-globin gene complex on chromosome 11. *Proc. Natl. Acad. Sci. U.S.A.* **76**, 5239–5243.

Gutman, N., Rae, P., and Schimmer, B. (1978). Altered cyclic AMP-dependent protein kinase activity in a mutant adrenocortical tumor cell line. *J. Cell Physiol.* **97**, 451–460.

Haga, T., Ross, E., Anderson, H., and Gilman, A. (1977). Adenylate cyclase permanently uncoupled from hormone receptors in a novel variant from mouse lymphoma cells. *Proc. Natl. Acad. Sci. U.S.A.* **74**, 2016–2020.

Hakala, M. T. (1957). Prevention of toxicity of amethopterin for sarcoma 180 cells in tissue culture. *Science* **126**, 255–256.

Halaban, R., Norklund, J., Francke, U., Moellmann, G., and Eisenstadt, J. M. (1980). Supermelanotic hybrids derived from mouse melanomas and normal mouse cells. *Somatic Cell Genet.* **6**, 29–44.

Hamilton, T., Tin, A., and Sussman, H. (1979). Regulation of alkaline phosphatase in human choriocarcinoma cell lines. *Proc. Natl. Acad. Sci. U.S.A.* **76**, 323–327.

Hankinson, O. (1976). Mutants of the Chinese hamster ovary cell line requring alanine and glutamate. *Somatic Cell Genet.* **2**(6), 497–507.

Hankinson, O. (1979). Sinole-step selection of clones of a mouse hepatoma line deficient in aryl hydrocarbon hydroxylase. *Proc. Natl. Acad. Sci. U.S.A.* **76**(1) 373–376.

Harris, H. (1967). The reactivation of the red cell nucleus. *J. Cell Sci.* **2**, 23–32.

Harris, H. (1970). "Cell Fusion." Harvard Univ. Press, Cambridge, Massachusetts.

Harris, H. (1971). The Croonian lecture: Cell fusion and the analysis of malignancy. *Proc. R. Soc. London, Ser. B* **179**, 1–20.

Harris, H. (1979). Some recent progress in the analysis of malignancy by cell fusion. *Ciba Found. Symp.* **66**, (new ser.), 311–333.

Harris, H. (1980). Switching on the muscle genes. *Nature (London)* **286**, 758–759.

Harris, H., and Klein, G. (1969). Malignancy of somatic. cell hybrids. *Nature (London)* **224**, 1315–1316.

Harris, M. (1964). "Cell Culture and Somatic Variation." Holt, Rinehart & Winston, New York.

Harris, M. (1971). Mutation rates in cells at different ploidy levels. *J. Cell. Physiol.* **78**, 177–184.

Harris, M. (1978). Cytoplasmic transfer of resistance to antimycin A in Chinese hamster cells. *Proc. Natl. Acad. Sci. U.S.A.* **75**, 5604–5608.

Harris, M. (1982). Induction of thymidine kinase in enzyme-deficient Chinese hamster cells. *Cell* **29**, 483–492.

Harris, M., and Collier, K. (1980). Phenotype evolution of cells resistant to bromodeoxyuridine. *Proc. Natl. Acad. Sci. U.S.A.* **77**(7), 4206–4210.

Harrison, D. E. (1975). Normal function of transplanted marrow cell lines from aged mice. *J. Gerontol.* **30**, 279–285.

Hay, R., Menzies, R., Morgan, H., and Strehler, B. (1968). The division potential of cells in continuous growth as compared to cells subcultivated after maintenance in stationary phase. *Exp. Gerontol.* **3**, 35–44.

Hayflick, L. (1965). The limited *in vitro* lifetime of human diploid cell strains. *Exp. Cell Res.* **37**, 614–636.

Hayflick, L. (1968). Human cells and aging. *Sci. Am.* **218**, 21–37.

Hayflick, L. (1973). The biology of human aging. *Am. J. Med. Sci.* **265**, 432–445.

Hayflick, L. (1979). The cell biology of aging. *J. Ivest. Dermatol.* **73**, 8–14.

Hayflick, L. (1980). The cell biology of human aging. *Sci. Am.* **242**, 58–66.

Hazum, E., Chang, K.-J., and Cuatrecasas, P. (1979). Opiate (enkephalin) receptors of neuroblastoma cells: Occurrence in clusters on the cell surface. *Science* **206**, 1077–1079.

Henle, W., Henle, G., and Lennette, E. T. (1979). The Epstein–Barr virus. *Sci. Am.* **241**, 48–59.

Hertz, F. (1973). Alkaline phosphatase in KB cells: Influence of hyperosmolality and prednisilone on enzyme activity and thermostability. *Arch. Biochem. Biophys.* **158**, 225–235.

Hill, J. M., Roberts, J., Loeb, E., Khan, A., MacLellan, A., and Hill, R. B. (1967). L-asparaginase therapy for leukemia and other malignant neoplasms. *JAMA, J. Am. Med. Assoc.* **202**, 882–888.

Hitotsumachi, S., Rabinowitz, Z., and Sachs, L. (1971). Chromosomal control of reversion in transformed cells. *Nature (London)* **231**, 511–514.

Hoar, D., and Sargent, P. (1976). Chemical mutagen hypersensitivity in ataxia telangectasia. *Nature (London)* **261**, 590–592.

Holland, J. J., Kohne, D., and Doyle, M. F. (1973). Analysis of virus replication in aging human fibroblast cultures. *Nature (London)* **245**, 209–214.

Holley, R. (1975). Control of growth in mammalian cells in culture. *Nature (London)* **258**, 487–490.

Holliday, R. (1975). Testing the protein error theory of aging: A reply to Baird, Samis, Massie and Zimmerman. *Gerontologia* **21**, 64–68.

Holliday, R., and Tarrant, G. M. (1972). Altered enzymes in aging human fibroblasts. *Nature (London)* **238**, 26–30.

Holtzer, H., Weintraub, H., Mayne, R., and Mochran, B. (1972). The cell cycle, cell lineages and cell differentiation. *Curr. Top. Dev. Biol.* **7**, 229–256.

Howell, A., and Sager, R. (1978). Tumorigenicity and its suppression and cybrids of mouse and Chinese hamster cell lines. *Proc. Natl. Acad. Sci. U.S.A.* **75**, 2358–2362.

Hozier, J., Sawyer, J., Moore, M., Howard, B., Clive, D. (1981). Cytogenetic analysis of the L5178Y/TK$^{+/-} \rightarrow$ TK$^{-/-}$ mouse lymphoma mutagenesis assay system. *Mutat. Res.* **84**, 169–181.

Hsie, A. (1981). Reverse transformation of Chinese hamster cells by cyclic AMP and hormones. *In* "Cell Growth" (C. Nicolini, ed.), pp. 557–574. Plenum, New York.

Hsie, A., and Puck, T. (1971). Morphological transformation of Chinese hamster ovary cells by dibutyryl adenosine cyclic $3':5'$-monophosphate and testosterone. *Proc. Natl. Acad. Sci. U.S.A.* **68**, 358–361.

Hsie, A., Brimer, P., Mitchell, T., and Gossilee, D. (1975). The dose–response relationship for ultraviolet-light-induced mutations at the hypoxanthine–guanine phosphoribosyltransferase locus in Chinese hamster ovary cells. *Somatic Cell Genet.* **1**, 383–389.

Hsiung, N., Warrick, H., DeRiel, J., Tuan, D., Forget, B., Skoultchi, A., and Kucherlapati, R. (1980). Cotransfer of circular and linear prokaryotic and eukaryotic DNA sequences into mouse cells. *Proc. Natl. Acad. Sci. U.S.A.* **77**, 4852–4856.

Huberman, E., and Sachs, L. (1974). Cell mediated mutagenesis of mammalian cells with chemical carcinogens. *Int. J. Cancer* **13**, 326–333.

Huberman, E., and Sachs, L. (1976). Mutability of different genetic loci in mammalian cells by metabolically activated carcinogenic polycyclic hydrocarbons. *Proc. Natl. Acad. Sci. U.S.A.* **73**, 188–192.

Huberman, E., Sachs, L., Yang, S., and Gelboin, H. (1976). Identification of mutagenic metabolites of benzo[*a*]pyrene in mammalian cell. *Proc. Natl. Acad. Sci. U.S.A.* **73**, 607–611.

Ikawa, Y., Kameji, R., Uchiyama, Y., Inoue, Y., Aida, M., and Obinata, M. (1979). The Friend leukemia cell system as a model for the molecular genetics of erythrodifferentiation. *In* "Mechanisms of Cell Change" (J. Ebert and T. Okada, eds.), pp. 83–98. Wiley, New York.

Illmensee, K., and Stevens, L. C. (1979). Teratomas and chimeras. *Sci. Am.* **240**, 87–97.

Ingles, C. J. (1978). Temperature-sensitive RNA polymerase II mutations in Chinese ovary cells. *Proc. Natl. Acad. Sci. U.S.A.* **75**(1), 405–409.

Ingles, C. J., Guiala, A., Lam, J., and Siminovitch, L. (1976). Alpha-amanitin resistance of RNA polymerase II in mutant Chinese hamster ovary cell lines. *J. Biol. Chem.* **251**, 2729–2734.

Insley, J., Bird, G., Harper, P. S., and Pearce, G. (1976). Prenatal prediction of myotonic dystrophy. *Lancet* **1**, 806–807.

Irwin, M., Oates, D. C., and Patterson, D. (1979). Biochemical genetics of Chinese hamster cell mutants with deviant purine metabolism: Isolation and characterization of a mutant deficient in the activity of phosphoribosylaminoimidazole synthetase. *Somatic Cell Genet.* **5**(2), 203–216.

Jacob, F., and Monod, J. (1961). Genetic regulatory mechanisms in the synthesis of protein. *J. Mol. Biol.* **3**, 318–356.

Jacobson, C. (1969). Reactivation of DNA synthesis in mammalian neuron nuclei after fusion with cells of an undifferentiated fibroblast line. *Exp. Cell Res.* **53**, 316–318.

Jacobson, K., Krell, M., Dempsey, M., Lugo, O., Ellingson, O., and Hench, C. (1978). Toxicity and mutagenicity of radiation from fluorescent lamps and a sunlamp in L5178y mouse lymphoma cells. *Mutat. Res.* **51**, 61–75.

Jami, J., and Ritz, E. (1973). Non-malignancy hybrids derived from two mouse malignant cell. I. Hybrids between L1210 leukemia cells and malignant L cells. *JNCI, J. Natl. Cancer Inst.* **51**, 1647–1653.

Jami, J., and Ritz, E. (1975). Non-malignancy of hybrids derived from two mouse malignant cells. II. Analysis of malignancy of LM(TK$^-$) Cl 1D parental cells. *JNCI, J. Natl. Cancer Inst.* **54**, 117–122.

Jonasson, J., and Harris, H. (1977). The analysis of malignancy by cell fusion. VIII. Evidence for the intervention of an extra-chromosomal element. *J. Cell Sci.* **24**, 255–263.

Jones, C., and Moore, E. E. (1976). Isolation of mutants lacking branched-chain amino acid transaminases. *Somatic Cell Genet.* **2**, 235–243.

Jones, C. W., Mastrangelo, I. A., Smith, H. H., Liu, H. Z., and Meck, R. A. (1976). Interkingdom fusion between human (HeLa) cells and tobacco hybrid (GGLL) protoplasts. *Science* **193**, 401–403.

Jones, G. E., and Sargent, P. (1974). Mutants of cultured Chinese hamster cells deficient in adenine phosphoribosyl transferase. *Cell* **2**, 43–54.

Jones, K. (1980). Muscle cell differentiation and the prospects for genetic engineering. *Br. Med. Bull.* **36**, 173–180.

Jones, P., and Taylor, S. (1980). Cellular differentiation, cytidine analogs and DNA methylation. *Cell* **20**, 85–93.

Jostes, R. F., Dewey, W. C., and Hopwood, L. E. (1977). Mutagenesis by fluorescent light in mammalian cell cultures. *Mutat. Res.* **42**, 139–144.

Kahan, B., and DeMars, R. (1975). Localized derepression on the human inactive X chromosome in mouse–human cell hybrids. *Proc. Natl. Acad. Sci. U.S.A.* **72**, 1510–1514.

Kahan, B., and DeMars, R. (1980). Autonomous gene expression on the human inactive X chromosomes. *Somatic Cell Genet.* **6**, 309–323.

Kahn, A., Guillouzo, A., Liebovitch, M., Cottreau, D., Bourel, M., and Dreyfus, J. (1977). Heat lability of glucose-6-phosphate dehydrogenase in some senescent human cultured cells. Evidence for its post synthetic nature. *Biochem. Biophys. Res. Commun.* **77**, 760–766.

Kallos, J., Fasy, T., Hollander, V., and Bick, M. (1978). Estrogen receptor has enhanced affinity for bromodeoxyuridine-substituted DNA. *Proc. Natl. Acad. Sci. U.S.A.* **75**, 4896–4900.

Kan, Y., and Dozy, A. (1978). Polymorphism of DNA sequence adjacent to the human beta globin structural gene. Its relation to the sickle mutation. *Proc. Natl. Acad. Sci. U.S.A.* **75**, 5631–5635.

Kao, F. T. (1973). Identification of chick chromosomes in cell hybrids formed between chick erythrocytes and adenine-requiring mutants of Chinese hamster cells. *Proc. Natl. Acad. Sci. U.S.A.* **70**, 2893–2898.

Kao, F. T., and Puck, T. T. (1968). Genetics of somatic mammalian cells. VII. Induction and isolation of nutritional mutants in Chinese hamster cells. *Proc. Natl. Acad. Sci. U.S.A.* **60**, 1275–1281.

Kao, F. T., and Puck, T. T. (1969). Genetics of somatic mammalian cells. IX. Quantitation of mutagenesis by physical and chemical agents. *J. Cell. Physiol.* **74**, 245–258.

Kao, F. T., and Puck, T. T. (1971). Genetics of somatic mammalian cells. XII. Mutagenesis by carcinogenic nitroso compounds. *J. Cell. Physiol.* **78**, 139–144.

Kaplan, M., and Koprowski, H. (1980). Rabies. *Sci. Am.* **242**(1), 120–138.

Kaufman, S., and Parks, C. (1977). Loss of growth control and differentiation in the fu-1 variant of the L$_8$ line of rat myoblasts. *Proc. Natl. Acad. Sci. U.S.A.* **74**, 3888–3892.

Keenan, B., Meyer, W. J., Hadjian, A., Jones, H., and Migeon, C. (1974). Syndrome of androgen insensitivity in man: Absence of 5-alpha-dihydrotestosterone binding protein in skin fibroblasts. *J. Clin. Endocrinol. Metab.* **38**, 1143–1146.

Keenan, B., Meyer, W. J., Hadjian, A., and Migeon, C. (1975). Androgen receptors in human skin fibroblasts: Characterization of a specific 17-beta-hydroxy-5-alpha-androstan-3-one protein complex in cell sonicates and nuclei. *Steroids* **25**, 535–552.

Kempe, T. D., Swyryd, E. A., Bruist, M., and Stark, G. A. (1976). Stable mutants of mammalian cells that overproduce the first three enzymes of pyrimidine nucleotide biosynthesis. *Cell* **9**, 541–550.

Kenney, F., Lee, K., Stiles, C., and Fritz, J. (1973). Further evidence against posttranscriptional control of inducible tyrosine amino transferase synthesis in cultured hepatoma cells. *Nature (New Biol.)* **246**, 208–210.

Kenney, F., Lee, K., Pomato, N., and Nickol, J. (1979). Multiple hormonal control of enzyme synthesis in liver and hepatoma cells. *In* "Hormones and Cell Culture" (G. Sato and R. Rose, eds.), pp. 905–917. Cold Spring Harbor Lab., Cold Spring Harbor, New York.

Kit, S., and Leung, W. C. (1974). Genetic control of mitochondrial thymidine kinase in human–mouse and monkey–mouse somatic cell hybrids. *J. Cell Biol.* **61**, 35–44.

Kit, S., Dubbs, D. R., Piekarski, L. J., and Hsu, T. C. (1963). Deletion of thymidine kinase activity from L cells resistant to bromodeoxyuridine. *Exp. Cell Res.* **31**, 297–312.

Klee, W., and Nirenberg, M. (1974). A neuroblastoma × glioma hybrid cell line with morphine receptors. *Proc. Natl. Acad. Sci. U.S.A.* **71**, 3474–3477.

Klobutcher, L. A., and Ruddle, F. H. (1979). Phenotype stabilization and integration of transferred materials in chromosome mediated gene transfer. *Nature (London)* **280**, 657–660.

Knudson, A. G. (1977). Mutation and cancer in man. *Cancer* **39**, 1882–1886.

Köhler, G., and Milstein, C. (1975). Continuous cultures of fused cells secreting antibody of predefined specificity. *Nature (London)* **256**, 595–597.

Köhler, G., and Milstein, C. (1976). Deviation of specific antibody-producing tissue culture and tumor lines by cell fusion. *Eur. J. Immunol.* **6**, 511–519.

Kohn, R. R. (1975). Aging and cell division. *Science* **188**, 203–204.

Kolata, G., and Wade, N. (1980). Human gene treatment stirs new debate. *Science* **210**, 407.

Koprowski, H., Gerhard, W., and Croce, C. (1977). Production of antibodies against influenza virus by somatic cell hybrids between mouse myeloma and primed spleen cells. *Proc. Natl. Acad. Sci. U.S.A.* **74**, 2985–2988.

Koprowski, H., Steplewski, Z., Herlyn, D., and Herlyn, M. (1978). Study of antibodies against human melanoma produced by somatic cell hybrids. *Proc. Natl. Acad. Sci. U.S.A.* **75**, 3405–3409.

Krag, S. S. (1979). A concanavalin A-resistant Chinese hamster ovary cell line is deficient in the synthesis of [^{3}H] glucosyl oligosaccharide lipid. *J. Biol. Chem.* **254**, 9167–9177.

Krieger, N., Brown, M., and Goldstein, J. (1981). Isolation of Chinese hamster cell mutants defective in the receptor-mediated endocytosis of low density lipoprotein. *J. Mol. Biol.* **150**, 167–184.

Krohn, P. (1962). Heterochromic transplantation in the study of aging. *Proc. R. Soc. London, Ser. B* **157**, 128–147.

Krohn, P. (1965). *In* "Advances in the Biology of Skin" (W. Montagna, ed.), Vol. 6. Pergamon, Oxford.

Krooth, R. S., Hsiao, W. L., Potvin, B. W. (1979). Resistance to 5-fluoroorotic acid and pyrimidine auxotrophy: A new bidirectional selective system for mammalian cells. *Somatic Cell Genet.* **5**, 551–569.

Kuby, S., Hans, H. J., Okabe, K., Jacobs, H. K., Ziter, F., Gerber, D., and Tyler, F. H. (1977).

Isolation of the human ATP-creatine transphosphorylases (creatine phosphokinases) from tissues of patients with Duchenne muscular dystrophy. *J. Biol. Chem.* **252**, 8382–8390.

Kucherlapati, R. S., and Ruddle, F. H. (1976). Advances in human gene mapping by parasexual procedures. *Prog. Med. Genet.* **1**, 121–144.

Kucherlapati, R. S., and Shin, S. (1979). Genetic control of tumorigenicity in interspecific mammalian cell hybrids. *Cell* **16**, 639–648.

Lai, E., Woo, S., Bordelon-Riser, M., Fraser, T., and O'Malley, B. (1980). Ovalbumin is synthesized in mouse cells transformed with the natural chicken ovalbumin gene. *Proc. Natl. Acad. Sci. U.S.A.* **77**, 244–248.

Law, L. W. (1952). Origin of the resistance of leukemia cells to folic-acid antagonists. *Nature (London)* **169**, 628–629.

Lea, D. E., and Coulson, C. A. (1949). The distribution of the numbers of mutants in a bacterial population. *J. Genet.* **49**, 264–285.

LeCam, A., Nicolas, J., Singh, T., Cabral, F., Pastan, I., and Gottesman, M. (1981). Cyclic AMP-dependent phosphorylation in intact cells and in cell-free extracts from Chinese hamster ovary cells. *J. Biol. Chem.* **256**, 933–941.

Leder, A., and Leder, P. (1975). Butyric acid, a potent inducer of erythroid differentiation in cultured erythroleukemia cells. *Cell* **5**, 319–322.

Leder, A., Swan, D., Ruddle, F., D'Eustachio, P., and Leder, P. (1981). Dispersion of alpha-like globin genes of the mouse to three different chromosomes. *Nature (London)* **293**, 196–200.

Lederberg, J., and Lederberg, E. M. (1952). Replica plating and indirect selection of bacterial mutants. *J. Bacteriol.* **63**, 399–406.

Leinwand, L., Strair, R., and Ruddle, F. H. (1978). Phenotypic and molecular expression of albumin in rat hepatoma × L cell hybrids. *Exp. Cell Res.* **115**, 261–268.

Lennon, V., and Lambert, E. (1980). Myasthenia gravis induced by monoclonal antibodies to acetylcholine receptors. *Nature (London)* **285**, 238–240.

Levisohn, S. R., and Thompson, E. B. (1973). Contact inhibition and gene expression in HTC/L cell hybrid lines. *J. Cell. Physiol.* **81**, 225–232.

Levitt, D., and Dorfman, A. (1972). The irreversible inhibition of differentiation of limb-bud mesenchyme by bromodeoxyuridine. *Proc. Natl. Acad. Sci. U.S.A.* **69**, 1253–1257.

Levy, J., Terada, M., Rifkind, R., and Marks, P. (1975). Induction of erythroid differentiation by dimethylsulfoxide in cells infected with Friend virus: Relationship to the cell cycle. *Proc. Natl. Acad. Sci. U.S.A.* **72**, 28–32.

Lewis, C. M., and Tarrant, G. M. (1972). Error theory and aging in human diploid fibroblasts. *Nature (London)* **239**, 316–318.

Lieberman, I., and Ove, P. (1959). Estimation of mutation rates with mammalian cells in culture. *Proc. Natl. Acad. Sci. U.S.A.* **45**, 872–877.

Lin, S., and Riggs, A. (1972). *Lac* operator analogues: Bromodeoxyuridine substitution in the *lac* operator affects the rate of dissociation of the *lac* repressor. *Proc. Natl. Acad. Sci. U.S.A.* **69**, 2574–2576.

Lin, S., and Riggs, A. (1976). The binding of *lac* repressor and the catabolite gene activator protein to halogen-substituted analogues of poly[d(A–T)]. *Biochim. Biophys. Acta* **432**, 185–191.

Linder, D., and Gartler, S. M. (1965). Glucose-6-phosphate dehydrogenase mosaicism: Utilization as a cell marker in the study of leiomyomas. *Science* **150**, 67–69.

Ling, V. (1975). Drug resistance and membrane alteration in mutants of mammalian cells. *Can. J. Genet. Cytol.* **17**, 503–515.

Ling, V., and Baker, R. M. (1978). Dominance of colchicine resistance in hybrid CHO cells. *Somatic Cell Genet.* **4**, 193–200.

Ling, V., Aubin, J. E., Chase, A., and Sarangi, F. (1979). Mutants of Chinese hamstery ovary (CHO) cells with altered colcemid-binding affinity. *Cell* **18**, 423–430.

Linn, S., Kairis, M., and Holliday, R. (1976). Decreased fidelity of DNA polymerase activity isolated from aging human fibroblasts. *Proc. Natl. Acad. Sci. U.S.A.* **73,** 2818–3822.

Lipsich, L. A., Kates, J. R., and Lucas, J. J. (1979). Expression of a liver-specific function by mouse fibroblast nuclei transplanted into rat hepatoma cytoplasts. *Nature (London)* **281,** 74–76.

Liskay, R. M., and Patterson, D. (1979). A selective medium (GAMA) for the isolation of somatic cell hybrids from HPRT⁻ and APRT⁻ mutant cells. *Cytogenet. Cell Genet.* **23,** 61–69.

Littlefield, J. W. (1964). Selection of hybrids from matings of fibroblasts *in vitro* and their presumed recombinants. *Science* **145,** 709–710.

Littlefield, J. W. (1965). Studies on thymidine kinase in cultured mouse fibroblasts. *Biochim. Biophys. Acta* **95,** 14–22.

Littlefield, J. W. (1969). Hybridization of hamster cells with high and low folate reductase activity. *Proc. Natl. Acad. Sci. U.S.A.* **62,** 88–95.

Liu, C., Slate, D., Gravel, R., and Ruddle, F. (1979). Biological detection of specific mRNA molecules by microinjection. *Proc. Natl. Acad. Sci. U.S.A.* **76,** 4503–4506.

Loftfield, R. (1963). The frequency of errors in protein biosynthesis. *Biochem. J.* **128,** 1353–1356.

Loftfield, R., and Vanderjagt, D. (1972). The frequency of errors in protein biosynthesis. *Biochem. J.* **128,** 1353–1356.

Lucchesi, J. C. (1978). Gene dosage compensation and the evolution of sex chromosomes. *Science* **202,** 711–716.

Luk, D., and Bick, M. (1977). Determination of 5′-bromodeoxyuridine in DNA by buoyant density. *Anal. Biochem.* **77,** 346–349.

Luria, S., and Delbrück, M. (1943). Mutations of bacteria from virus sensitivity to virus resistance. *Genetics* **28,** 491–511.

Lyon, M. F. (1961). Gene action in the X chromosome of the mouse. *Nature (London)* **190,** 372–373.

Lyon, M. F. (1972). X-chromosomal inactivation and developmental patterns in mammals. *Biol. Rev. Cambridge Philos. Soc.* **47,** 1–36.

McBride, O. W., and Athwal, R. S. (1976). Genetic analysis by chromosome-mediated transfer. *In Vitro* **12,** 777–786.

McBride, O. W., and Ozer, H. L. (1973). Transfer of genetic information by purified metaphase chromosomes. *Proc. Natl. Acad. Sci. U.S.A.* **70,** 1258–1262.

McBride, O. W., Burch, J. W., and Ruddle, F. H. (1978). Cotransfer of thymidine kinase and galactokinase genes by chromosome-mediated gene transfer. *Proc. Natl. Acad. Sci. U.S.A.* **75,** 914–918.

McBurney, M., Featherstone, M., and Kaplan, H. (1978). Activation of teratocarcinoma-derived hemoglobin genes in teratocarcinoma–Friend cell hybrids. *Cell* **15,** 1323–1330.

McBurney, M. W., and Whitmore, G. F. (1974). Isolation and biochemical characterization of folate deficient mutants of Chinese hamster cells. *Cell* **2,** 173–182.

McCann, J., and Ames, B. N. (1976). Detection of carcinogens as mutagens in the Salmonella microsome test: Assay of 300 chemicals: Discussion. *Proc. Natl. Acad. Sci. U.S.A.* **73,** 950–954.

McCurdy, P. (1971). Use of genetic linkage for the detection of female carriers of hemophilia. *N. Engl. J. Med.* **285,** 218–219.

McGhee, J., and Ginder, G. (1979). Specific DNA methylation sites in the vicinity of the chicken beta-globin genes. *Nature (London)* **280,** 419–420.

McGrath, M., Pillemer, E., and Weissman, I. (1980). Murine leukaemogenesis: Monoclonal antibodies to T-cell determinants arrest T-lymphoma cell proliferation. *Nature (London)* **285,** 259–261.

McHale, J. S., Moulton, M. L., and McHale, J. T. (1971). Limited culture life span of human diploid cells as a function of metabolic time instead of division potential. *Exp. Gerontol.* **6,** 89–93.

Macieira-Coelho, A. (1980). Implications of the reorganization of the cell genome for aging or immortalization of dividing cells *in vitro. Gerontology* **26,** 276–282.

McKearn, T., Fitch, F., Smilek, D., Sarmiento, M., and Stuart, F. (1979). Properties of rat anti-MHC antibodies produced by cloned rat–mouse hybridomas. *Immunol. Rev.* **47,** 91–116.

McKusick, V. (1969). "Human Genetics." Prentice-Hall, Englewood Cliffs, New Jersey.

McKusick, V., and Ruddle, F. H. (1977). The status of the gene map of the human chromosomes. *Science* **196,** 390–405.

McKusick, V. A. (1980). The anatomy of the human genome. *J. Hered.* **71,** 370–391.

McKusick, V. A. (1982). The human genome through the eyes of a clinical geneticist. *Cytogenet. Cell Genet.* **32,** 7–23.

McMorris, E., Kolber, A., Moore, B., and Perumal, A. (1974). Expression of the neuron-specific protein, 14-3-2, and steroid sulfatase in neuroblatoma cell hybrids. *J. Cell Physiol.* **84,** 473–480.

Maher, V., Ouellette, L., Curren, R. D., and McCormick, J. (1976). Frequency of ultraviolet induced mutations is higher in xeroderma pigmentosum variant cells than in normal human cells. *Nature (London)* **261,** 593–595.

Malawista, S. E., and Weiss, M. C. (1974). Expression of differentiated functions in hepatoma cell hybrids: High frequency of induction of mouse albumin production in rat hepatoma–mouse lymphoblast hybrids. *Proc. Natl. Acad. Sci. U.S.A.* **71,** 927–931.

Malling, H. (1971). Dimethylnitrosamine formation of mutagenic compounds by interaction with mouse liver microsomes. *Mutat. Res.* **13,** 425–429.

Martin, G. W., Sprague, C. A., and Epstein, C. J. (1970). Replicative life span of cultivated human cells effect of donor's age, tissue and genotype. *Lab. Invest.* **23,** 86–92.

Marx, J. (1981). Antibodies: Getting their genes together. *Science* **212,** 1015–1017.

Masui, H., Reid, L. M., and Glans, C. (1978). Isolation and characterization of cyclic AMP-resistant variants from a functional adrenal cortical cell line. *Exp. Cell Res.* **117,** 219–230.

Mathis, A., and Fischer, G. A. (1962). Transformation experiments with murine lymphoblastic cells grown in culture. *Biochem. Pharmacol.* **11,** 69–78.

Mattern, M., and Cerutti, P. (1975). Age dependent excision repair of damaged thymine from gamma-irradiated DNA by isolated nuclei from human fibroblasts. *Nature (London)* **254,** 450–452.

Maurer, R. T., Kaji, H., and Morrow, J. (1983). Suppression of immunoglobulin synthesis in somatic cell hybrids between polyploid myeloma cells and mouse fibroblasts. (In preparation.)

Medrano, L., and Green, H. (1974). A uridine kinase-deficient mutant of 3T3 and a selective method for cells containing the enzyme. *Cell* **1,** 23–26.

Meo, T., Johnson, J., Beechy, C., Andrews, S., Peters, J., and Searle, A. (1980). Linkage analysis of murine immunoglobulin heavy chain and serum prealbumin genes establish their location on chromosome 12 proximal to the *T95;12)31H* breakpoint in band 12 FL. *Proc. Natl. Acad. Sci. U.S.A.* **77,** 550–553.

Merrill, G., Witter, E., and Hauschka, S. (1980). Differentiation of thymidine kinase deficient mouse myoblasts in the presence of 5' bromodeoxyuridine. *Exp. Cell Res.* **129,** 191–199.

Mertz, J. E., and Gurdon, J. (1977). Purified DNA's are transcribed after microinjection into *Xenopus* oocytes. *Proc. Natl. Acad. Sci. U.S.A.* **74,** 1502–1506.

Meyer, W., Migeon, B., and Migeon, C. (1975). Locus on human X chromosome for dihydro-testosterone receptors and androgen insensitivity. *Proc. Natl. Acad. Sci. U.S.A.* **72,** 1469–1472.

Meyers, M. B., van Diggelen, O. P., van Diggelen, M., and Shin, S. (1980). Isolation of somatic cell mutants with specified alterations in hypoxanthine phosphoribosyltransferase. *Somatic Cell Genet.* **6,** 290–306.

Migeon, B. E., der Kaloustian, V. M., Nyhan, W. L., Young, W. J., and Childs, B. (1968). X-linked hypoxanthine–guanine phosphoribosyltransferase deficiency: Heterozygote has two clonal populations. *Science* **160**, 425–426.

Miller, C. L., and Ruddle, F. H. (1978). Co-transfer of human X-linked markers into murine somatic cells via isolated metaphase chromosomes. *Proc. Natl. Acad. Sci. U.S.A.* **75**, 3346–3350.

Milman, G., Krauss, S. W., and Olsen, A. S. (1977). Tryptic peptide analysis of normal and mutant forms of hypoxanthine phosphoribosyltransferase from HeLa cells. *Proc. Natl. Acad. Sci. U.S.A.* **74**, 926–930.

Milstein, C., Adetugbo, K., Cowan, N., Kohler, G., Secher, D., and Wilde, C. (1977). Somatic cell genetics of antibody secreting cells—studies of clonal diversification and analysis by cell fusion. *Cold Spring Harbor Symp. Quant. Biol.* **4**, 793–803.

Minna, J., Nelson, P., Peacock, J. (1971). Genes for neuronal properties expressed in neuroblastoma × L cell hybrids. *Proc. Natl. Acad. Sci. U.S.A.* **68**, 234–239.

Minna, J., Glaser, D., Nirenberg, M. (1972). Genetic dissection of neural properties using somatic cell hybrids. *Nature (London), New Biol.* **235**, 225–231.

Minna, J. D., and Coon, H. G. (1974). Human × mouse hybrid cells segregating mouse chromosomes and isozymes. *Nature (London)* **252**, 401–404.

Mintz, B., and Illmensee, K. (1975). Normal genetically mosaic mice produced from malignant teratocarcinoma cells. *Proc. Natl. Acad. Sci. U.S.A.* **72**, 3585–3589.

Mitchell, C. H., England, J. M., and Attardi, G. (1975). Isolation of chloramphenicol-resistant variants from a human cell line. *Somatic Cell Genet.* **1**, 215–234.

Moehring, J. M., Moehring, T. J., and Danley, D. E. (1980). Posttranslational modification of elongation factor 2 in diphtheria-toxin-resistant mutants of CHO-K1 cells. *Proc. Natl. Acad. Sci. U.S.A.* **77**, 1010–1014.

Mohandas, T., Sparkes, R. S., Hellkuhl, B., Grzeschik, K. H., and Shapiro, L. J. (1980). Expression of an X-linked gene from an inactive human X chromosome in mouse–human hybrid cells: Further evidence for the non-inactivation of the steroid sulfatase locus in man. *Proc. Natl. Acad. Sci. U.S.A.* **77**, 6759–6763.

Mohandas, T., Sparkes, R. S., and Shapiro, L. S. (1981). Reactivation of an inactive X chromosome: Evidence for X inactivation by DNA methylation. *Science* **211**, 393–396.

Molgaard, H. (1980). Assembly of immunoglobulin heavy chain genes. *Nature (London)* **286**, 657–659.

Morrow, J. (1964). Ph.d. dissertation.

Morrow, J. (1970). Genetic analysis of azaguanine resistance in an established mouse cell line. *Genetics* **65**, 279–287.

Morrow, J. (1971). Mutation rate from asparagine requirement to asparagine non-requirement. *J. Cell Physiol.* **77**, 423–425.

Morrow, J. (1972). Population dynamics of purine and pyrimidine sensitive and resistant cells grown in culture. *Genetics* **71**, 429–438.

Morrow, J. (1975). On the relationship between spontaneous mutation rates *in vivo* and *in vitro*. *Mutat. Res.* **33**, 367–372.

Morrow, J. (1977). Gene inactivation as a mechanism for the generation of variability in somatic cells cultivated *in vitro*. *Mutat. Res.* **44**, 391–400.

Morrow, J., Colofiore, J., and Rintoul, D. (1973). Azaguanine resistant hamster cell lines not deficient in hypoxanthine–guanine phosphoribosyl transferase. *J. Cell. Physiol.* **81**, 97–100.

Morrow, J., Prickett, M. S., Fritz, S., Vernick, D., and Deen, D. (1976). Mutagenesis studies on cultured mammalian cells. The sensitivity of the asparagine requiring phenotype of several chemical agents. *Mutat. Res.* **34**, 481–488.

Morrow, J., Stocco, D., and Barron, E. (1978). Spontaneous mutation rate to thioguanine resistance is decreased in polyploid hamster cell. *J. Cell. Physiol.* **96**, 81–86.

Morrow, J., Stocco, D., and Fralick, J. A. (1979). The requirement of DNA synthesis for the induction of alkaline phosphatase by bromodeoxyuridine in a derivative of the HeLa cell line. *J. Cell. Physiol.* **98**, 427–436.

Morrow, J., Sammons, D., and Barron, E. (1980). Puromycin resistance in Chinese hamster cells: Genetic and biochemical studies of partially resistant, unstable clones. *Mutat. Res.* **69**, 333–346.

Muggleton-Harris, A., and Palumbo, M. (1979). Nucleiocytoplasmic interactions in experimental binucleates formed from normal and transformed components. *Somatic Cell Genet.* **5**, 397–407.

Mulivor, R., Plotkin, L., and Harris, H. (1978). Differential inhibition of the products of the human alkaline phosphatase loci. *Ann. Hum. Genet.* **42**, 1–13.

Muller, H. J. (1947). Evidence of the precision of genetic adaptation. *Harvey Lect.* **43**, 165.

Naylor, S. L., Busby, L. L., and Klebe, R. J. (1976). Biochemical selection systems for mammalian cells: The essential amino acids. *Somatic Cell Genet.* **2**, 93–111.

Neff, J. M., and Enders, J. F. (1968). Poliovirus replication and cytopathogenicity in monolayer hamster cell cultures fused with beta propiolactone inactivated sendae virus. *Proc. Soc. Exp. Biol. Med.* **127**, 260–267.

Nelson, J. A., Carpenter, J. W., Rose, L. M., and Adamsen, D. J. (1975). Mechanism of action of 6-thioguanine, 6-mercaptopurine and 8-azaguanine. *Cancer Res.* **35**, 2872–2878.

Nelson, P., Christian, C., and Nirenberg, M. (1976). Synapse formation between clonal neuroblastoma × glioma hybrid cells and striated muscle cells. *Proc. Natl. Acad. Sci. U.S.A.* **73**, 123–127.

Nelson, R., Ruffner, W., and Nirenberg, M. (1969). Neuronal tumor cells with excitable membranes grown *in vitro. Proc. Natl. Acad. Sci. U.S.A.* **64**, 1004–1010.

Neufeld, E. F. (1974). Mucopolysaccharidoses, the biochemical approach. *In* "Medical Genetics" (V. McKusick and R. Claiborne, eds.), pp. 141–148. HP Publishing Company, New York.

Newbold, R., and Brookes, P. (1976). Exceptional mutagenicity of a benzo[a]pyrene diol epoxide in cultured mammalian cells. *Nature (London)* **261**, 53–54.

Nienhuis, A., and Benz, E., Jr. (1977). Regulation of hemoglobin synthesis during the development of the red cell. *N. Engl. J. Med.* **297**, 1371–1381.

Niwa, A., Yamamoto, K., and Yasumura, Y. (1979). Establishment of rat hepatoma cell line which has ornithine carbamoyltransferase activity and grows continuously in arginine-deprived medium. *J. Cell. Physiol.* **98**, 177–184.

Norwood, T. H., Pendergrass, W., and Martin, G. M. (1975). Reinitiation of DNA synthesis of senescent human fibroblasts upon fusion with cells of unlimited growth potential. *J. Cell Biol.* **64**, 551–556.

Nudel, U., Salmon, J., Terada, M., Bank, A., Rifkind, R., and Marks, P. (1977). Differential effects of chemical inducers on expression of beta globin genes in murine erythroleukemia cells. *Proc. Natl. Acad. Sci. U.S.A.* **74**, 1100–1104.

Nunberg, J. H., Kaufman, R. J., Schimke, R. T., Urlaub, G., and Chasin, L. A. (1978). Amplified dihydrofolate reductase genes are localized to a homogenously staining region of a single chromosome in a methotrexate-resistant Chinese hamster ovary cell line. *Proc. Natl. Acad. Sci. U.S.A.* **75**, 5553–5556.

Oates, D. C., Vannais, D., and Patterson, D. (1980). A mutant of CHO-K1 cells deficient in two nonsequential steps of *de novo* purine biosynthesis. *Cell* **20**, 797–805.

Ohno, S. (1971). Simplicity of mammalian regulatory systems inferred by single gene determination of sex phenotype. *Nature (London)* **234**, 134–137.

Ohno, S. (1973). Conservation of ancient linkage groups in evolution and some insight into the genetic regulatory mechanism of X-inactivation. *Cold Spring Harbor Symp. Quant. Biol.* **38**, 155–164.

Okada, U. (1958). The fusion of Erlich's tumor cells caused by HVJ virus *in vitro. Biken J.* **1**, 103–110.

O'Malley, B., and Schrader, W. (1976). Receptors of steroid hormones. *Sci. Am.* **234**(2), 32–43.

O'Malley, B., Towle, H., and Schwartz, R. (1977). Regulation of gene expression in eukaryotes. *Annu. Rev. Genet.* **11**, 239–275.

O'Neill, J., and Hsie, A. (1979). Phenotype expression time of mutagen-induced 6-thioguanine resistance in Chinese hamster ovary cells (CHO/HGPRT system). *Mutat. Res.* **59**, 109–118.

O'Neill, J., Couch, D., Machanoff, R., San Sebastian, J., Brimer, P., and Hsie, A. (1977). A quantitative assay of mutation induction at the hypoxanthine–guanine phosphoribosyl transferase locus in Chinese hamster ovary cells (CHO/HGPRT system): Utilization with a variety of mutagenic agents. *Mutat. Res.* **45**, 103–109.

Orci, L., Carpenter, J., Perrelet, A., Anderson, R., Goldstein, J. L., and Brown, M. S. (1978). Occurrence of low density lipoprotein receptors within large pits on the surface of human fibroblasts as demonstrated by freeze etching. *Exp. Cell Res.* **113**, 1–13.

Orgel, L. E. (1963). The maintenance of the accuracy of protein synthesis and its relevance to aging. *Proc. Natl. Acad. Sci. U.S.A.* **49**, 517–521.

Orgel, L. E. (1973). Aging of clones of mammalian cells. *Nature (London)* **243**, 441–445.

Orkin, S. (1978). The duplicated human alpha globin genes lie close together in cellular DNA. *Proc. Natl. Acad. Sci. U.S.A.* **75**, 5950–5954.

Orkin, S., Haroshi, F., and Leder, P. (1975). Differentiation in erythroleukemic cells and their somatic hybrids. *Proc. Natl. Acad. Sci. U.S.A.* **72**, 98–102.

Orkin, S. H., and Littlefield, J. W. (1971). Mutagenesis to aminopterin resistance in cultured hamster cells. *Exp. Cell Res.* **69**, 174–180.

Orkin, S. H., Buchanan, P. D., Yount, W. J., Reisner, H., and Littlefield, J. W. (1973). Lambda-chain production in human lymphoblast-mouse fibroblast hybrids. *Proc. Natl. Acad. Sci. U.S.A.* **70**, 2401–2405.

Ostertag, W., Crozier, T., Kluge, N., Melderis, H., and Dube, S. (1973). Action of 5-bromodeoxyuridine on the induction of haemoglobin synthesis in mouse leukemia cells resistant to 5-BUdR. *Nature (London), New Biol.* **243**, 203–205.

Ozer, H., and Jha, K. (1976). Malignancy and transformation: Expression in somatic cell hybrids and variants. *Adv. Cancer Res.* **25**, 53–89.

Palmiter, R. (1972). Regulation of protein synthesis in chick oviduct. *J. Biol. Chem.* **247**, 6770–6780.

Parsons, M., Morrow, J., Stocco, D., and Kitos, P. (1976). Purine uptake by azaguanine resistant hamster cells. *J. Cell. Physiol.* **89**, 209–218.

Passarge, E., and Fries, E. (1973). X-chromosomal inactivation in X-linked hyphohydrotic ectodermal dysplasia. *Nature (London), New Biol.* **245**, 58–59.

Patterson, D. (1980). Isolation and characterization of 5-fluorouracil resistant mutants of Chinese hamster ovary cells deficient in the activities or orotate phosphoribosyltransferase and orotidine 5′-monophosphate decarboxylase. *Somatic Cell Genet.* **610**, 101–114.

Patterson, D., and Carnwright, D. V. (1977). Biochemical genetic analysis of pyrimidine biosynthesis in mammalian cells: I. Isolation of a mutant defective in the early steps of *de novo* pyrimidine synthesis. *Somat. Cell Genet.* **3**, 483–495.

Patterson, M. K., and Orr, G. (1967). L-asparagine biosynthesis by nutritional variants of the Jensen sarcoma. *Biochem. Biophys. Res. Commun.* **26**, 228–233.

Patterson, D., Graw, S., and Jones, C. (1981). Demonstration using somatic cell genetics of coordinate regulation of genes for two enzymes of purine synthesis assigned to human chromosome 21. *Proc. Natl. Acad. Sci. U.S.A.* **78**, 405–409.

Pauling, L., Itano, H., Singer, S., Wells, I. (1949). Sickle cell anemia: A molecular disease. *Science* **110**, 543–544.

Periman, P. (1970). IgG synthesis in hybrid cells from an antibody-producing mouse myeloma and an L cell substrain. *Nature (London)* **228**, 1086–1087.

Perlow, M., Freed, W., Hoffer, B., Seiger, A., Olson, L., and Wyatt, R. (1979). Brain grafts reduce motor abnormalities produced by destruction of nigrostriatal dopamine system. *Science* **204**, 643–647.

Peterson, A. R., Krahn, D. F., Peterson, H., Heidelberger, C., Bhutan, B. K., and Li, L. H. (1976). The influence of serum components on the growth and mutation of Chinese hamster cells in medium containing 8-azaguanine. *Mutat. Res.* **36**, 345–356.

Peterson, A., Peterson, H., and Heidelberger, C. (1979). Oncogenesis, mutagenesis, DNA damage and cytotoxicity in cultured mammalian cells treated with alkylating agents. *Cancer Res.* **39**, 131–138.

Peterson, J. A. (1974). Discontinuous variability, in the form of a geometric progression, of albumin production in hepatoma and hybrid cells. *Proc. Natl. Acad. Sci. U.S.A.* **71**, 2062–2066.

Peterson, J. A. (1979). Analysis of discontinuous variation in albumin production by hepatoma cells at the cellular level. *Somatic Cell Genet.* **5**, 641–651.

Peterson, J. A., and Weiss, M. C. (1972). Expression of differentiated functions in hepatoma cell hybrids: Induction of mouse albumin production in rat hepatoma mouse fibroblast hybrids. *Proc. Natl. Acad. Sci. U.S.A.* **69**, 571–575.

Pfahl, M., and Bourgeois, S. (1980). Analysis of steroid resistance in lymphoma cell hybrids. *Somatic Cell Genet.* **6**, 63–74.

Pious, D., and Soderland, C. (1977). HLA variants of cultured human lymphoid cells: Evidence for mutational origin and estimation of mutation rates. *Science* **197**, 769–771.

Pöche, H., Varshaver, N., and Geissler, G. (1975). Cycloheximide resistance in Chinese hamster cells: I. Spontaneous mutagenesis. *Mutat. Res.* **27**, 399–406.

Pontecorvo, G. (1959). Panel discussion. *In* "Biochemistry of Human Genetics" (G. Wolsteinholme, C. O'Conner, J. Churchill, and A. Churchill, eds.), pp. 279–285. CIBA Foundation, London.

Pontecorvo, G. (1971). Induction of directional chromosome elimination in somatic cell hybrids. *Nature (London)* **230**, 367–369.

Pontecorvo, G. (1975). Symposium No. 19: Somatic cell genetics. Introduction by the Chairman. *Genetics* **79**(Suppl.), 339–341.

Prasad, K. (1971). Effect of dopamine and 6-hydroxydopamine on mouse neuroblastoma cells *in vitro. Cancer Res.* **31**, 1457–1460.

Prasad, K., and Mandel, B. (1973). Choline-acetyl transferase level in cyclic-AMP and x-ray induced morphologically differentiated neuroblastoma cells in culture. *Cytobiologie* **8**, 75–80.

Preisler, H., Housman, D., Scher, W., and Friend, C. (1973). Effects of 5-bromo-2-deoxyuridine on production of globin messenger RNA in dimethyl sulfoxide-stimulated friend leukemia cell. *Proc. Natl. Acad. Sci. U.S.A.* **70**, 2956–2959.

Prensky, W., and Holmquist, G. (1973). Chromosome localization of human haemoglobin structural genes: Techniques queried. *Nature (London)* **241**, 44–45.

Price, P., Conover, J., and Hirschen, K. (1972). Chromosomal localization of human hemoglobin structural genes. *Nature (London)* **237**, 340–342.

Prickett, M., Coultrip, L., Patterson, M., and Morrow, J. (1975). Effect of ploidy on spontaneous mutation rate to asparagine non-requirement in cultured cell. *J. Cell. Physiol.* **85**, 621–626.

Puck, T. T. (1975). Somatic cell genetics and its human applications. *Adv. Exp. Med. Biol.* **62**, 213–221.

Puck, T. T., Wuthier, P., Jones, C., and Kao, F. (1971). Genetics of somatic mammalian cells. XII. Lethal antigen as genetic markers for study of human linkage groups. *Proc. Natl. Acad. Sci. U.S.A.* **68**, 3102–3106.

Rabin, M. S., and Gottesman, M. M. (1979). High frequency of mutation to tubercidin resistance in CHO cells. *Somatic Cell Genet.* **5**, 571–583.

Rao, R. N., and Johnson, R. T. (1974). Regulation of cell cycle on hybrid cells. *Cold Spring Harbor Conf. Cell Proliferation* **1**, 785–800.

Raskind, W. H., and Gartler, S. M. (1978). The relationship between induced mutation frequency and chromosome dosage in established mouse fibroblast lines. *Somatic Cell Genet.* **4,** 491–506.

Rechsteiner, M., and Hill, D. (1975). Autoradiographic studies of nicotinic acid utilization in human–mouse heterokaryons and inhibition of utilization in newly formed hybrid cells. *J. Cell Physiol.* **86,** 439–452.

Reichenbecher, V. E., Jr., and Caskey, C. T. (1979). Emetine resistant Chinese hamster cells. *J. Biol. Chem.* **254,** 6207–6210.

Remsen, J., and Cerutti, P. (1976). Deficiency of gamma-ray excision repair in skin fibroblasts from patients with Fanconis anemia. *Proc. Natl. Acad. Sci. U.S.A.* **73,** 2419–2423.

Ricciuti, F. C., Gelehrter, T. D., and Rosenberg, L. E. (1976). X-chromosome inactivation in human liver: Confirmation of X-linkage of ornithine transcarbamylase. *Am. J. Hum. Genet.* **28,** 332–338.

Riddle, J., and Hsie, A. (1978). An effect of cell cycle position on ultraviolet-light-induced mutagenesis in Chinese hamster ovary cells. *Mutat. Res.* **52,** 409–420.

Riddle, V. G. H., and Harris, H. (1976). Synthesis of a liver enzyme in hybrid cells. *J. Cell Sci.* **22,** 199–215.

Riggs, A. (1975). X-inactivation differentiation and DNA methylation. *Cytogenet. Cell Genet.* **14,** 9–25.

Ringertz, N., Krondah, U., and Coleman, J. (1978). Reconstitution of cells by fusion of cell fragments. I. Myogenic expression after fusion of minicells from rat myoblasts (L6) with mouse fibroblast (A9) cytoplasm. *Exp. Cell Res.* **113,** 233–246.

Ringertz, N. R., and Savage, R. E. (1976). "Cell Hybrids." Academic Press, New York.

Rintoul, D., and Morrow, J. (1975). The use of selective markers in the study of differentiated gene function in normal and malignant cells of hepatic origin. *In* "Gene Expression and Carcinogenesis in Cultured Liver Cells" (L. E. Gerschenson and E. B. Thompson, eds.), pp. 311–324. Academic Press, New York.

Rintoul, D., Lewis, R. F., and Morrow, J. (1973a). Expression of differentiated functions in HTC–fibroblast hybrids. *Biochem. Genet.* **7,** 375–386.

Rintoul, D., Colofiore, J., and Morrow, J. (1973b). Expression of differentiated properties in fetal liver cells and their somatic cell hybrids. *Exp. Cell Res.* **78,** 414–422.

Ritz, J., Pesando, J., Notis-McConarty, J., Lazarus, H., and Schlossman, S. (1980). A monoclonal antibody to human acute lymphoblastic leukaemia antigen. *Nature (London)* **283,** 583–585.

Robbins, E., Levine, E. M., and Eagle, H. (1970). Morphological changes accompanying senescence of cultured human diploid cells. *J. Exp. Med.* **131,** 1211–1222.

Robertson, M. (1980). Chopping and changing in immunoglobin genes. *Nature (London)* **287,** 390–392.

Robertson, M. (1981). Genes of lymphocytes. I. Diverse means to antibody adversity. *Nature (London)* **290,** 625–627.

Robertson, M., and Hobart, M. (1981). Antibodies, introns and biosynthetic versatility. *Nature (London)* **290,** 5453–5444.

Rodbell, M., Lad, P., Nielsen, T., Cooper, D., Schlegel, W., Preston, M., Londos, C., and Kempner, E. (1981). The structure of adenylate cyclase systems. *Adv. Nucleotide Res.* **14,** 3–13.

Rogers, J., Ng, S. K., Coulter, M. B., and Sanwal, B. D. (1975). Inhibition of myogenesis in a rat myoblast line by 5-bromodeoxyuridine. *Nature (London)* **256,** 438–440.

Rogers, J., Coulter, M., Ng, S. K., and Sanwal, B. D. (1978). The noncoordinate expression of muscle-specific proteins in mutant rat skeletal myoblasts and reinitiation of differentiation in hybrids. *Somatic Cell Genet.* **4,** 573–585.

Romeo, G., and Migeon, B. (1970). Genetic inactivation of alpha galactosidase locus in carriers of Fabry's Disease. *Science* **170**, 180–181.

Rosen, R. (1978). Cells and senescence. *Int. Rev. Cytol.* **54**, 161–191.

Rosenberg, R. N. (1973). Tissue culture of the nervous system. *Curr. Top. Neurobiol.* **1**, 107–134.

Rosenberg, R. N., Vance, C., Morrison, M., Prashad, N., Meyne, J., and Baskin, F. (1978). Differentiation of neuroblastoma, glioma, and hybrid cells in culture as measured by the synthesis of specific protein species: Evidence for neuroblast–glioblast reciprocal genetic regulation. *J. Neurochem.* **30**, 1343–1355.

Rosenstraus, M. J., and Chasin, L. A. (1978). Separation of linked markers in Chinese hamster cell hybrids: Mitotic recombination is not involved. *Genetics* **90**, 735–760.

Ross, J., Aviv, H., Scolnick, E., and Leder, P. (1972). *In vitro* synthesis of DNA complementary to purified rabbit globin mRNA. *Proc. Natl. Acad. Sci. U.S.A.* **69**, 264–268.

Ross, R., and Sato, G. (1979). Cell culture and endocrinology: An overview. *In* "Hormones and Cell Culture" (G. Sato and R. Ross, eds.), pp. 965–971. Cold Spring Harbor Lab., Cold Spring Harbor, New York.

Rothstein, M. (1975). Aging and the alteration of enzymes: A review of mechanisms of aging and development. *Mech. Aging Dev.* **4**, 325–338.

Roufa, D. J., Sadow, B. N., and Caskey, C. T. (1973). Derivation of TK⁻ clones from revertant TK⁺ mammalian cells. *Genetics* **75**, 515–530.

Rousseau, G. G. (1975). Interaction of steroids with hepatoma cells: Molecular mechanisms of glucocorticoid hormone action. *J. Steroid Biochem.* **6**, 75–89.

Rubenstein, P. A., and Spudich, J. A. (1977). Actin microheterogeneity in chick embryo fibroblasts. *Proc. Natl. Acad. Sci. U.S.A.* **74**, 120–123.

Ruddle, F. H. (1981). A new era in mammalian gene mapping: Somatic cell genetics and recombinant DNA methodologies. *Nature (London)* **294**, 115–120.

Ruddle, F. H., and Kucherlapati, R. S. (1974). Hybrid cells and human genes. *Sci. Am.* **231**, 36–49.

Rutter, W., Pictet, R., and Morris, P. (1973). Toward molecular mechanisms of developmental processes. *Annu. Rev. Biochem.* **42**, 601–646.

Ryan, J. M., Duda, G., and Cristofalo, V. J. (1974). Error accumulation and aging in human diploid cells. *J. Gerontol.* **29**, 616–621.

Sabin, A. (1981). Suppression of malignancy in human cancer cells: Issues and challenges. *Proc. Natl. Acad. Sci. U.S.A.* **78**, 7129–7133.

Sachs, L. (1978). Control of normal cell differentiation and the phenotypic reversion of malignancy in myeloid leukemia. *Nature (London)* **274**, 535–539.

St. John, T., and Davis, R. (1979). Isolation of galactose-inducible DNA sequences from *Saccharomyces cerevisiae* by differential plaque filter hybridization. *Cell* **16**, 443–452.

Samaha, F., and Nagy, B. (1982). Distribution of protein components of the sarcoplasmic reticulum in normal and abnormal human muscle. *Ann. Neurol.* **12**, 108.

Sanwal, B. (1979). Myoblast differentiation. *Trends Biol. Sci.* **4**, 155–157.

Sato, K., and Heida, N. (1980). Mutation induction in a mouse lymphoma cell mutant sensitive to 4-nitroquinoline 1-oxide and ultraviolet radiation. *Mutat. Res.* **71**, 233–241.

Sato, K., Slesinski, R. S., and Littlefield, J. W. (1972). Chemical mutagenesis at the phosphoribosyltransferase locus in cultured human lymphocytes. *Proc. Natl. Acad. Sci. U.S.A.* **69**, 1244–1248.

Scaletta, L. J., and Ephrussi, B. (1965). Hybridization of normal and neoplastic cells *in vitro*. *Nature (London)* **205**, 1169–1171.

Scangos, G. A., Huitner, K. M., Silverstein, S., and Ruddle, F. H. (1979). Molecular analysis of chromosome mediated gene transfer. *Proc. Natl. Acad. Sci. U.S.A.* **76**, 3987–3990.

Schilling, J., Clevenger, B., Davie, J., and Hood, L. (1980). Amino acid sequence of homogeneous antibodies to dextran and DNA rearrangements in heavy chain V-region gene segments. *Nature (London)* **283**, 35–40.

Schimke, R. T. (1980). Gene amplification and drug resistance. *Sci. Am.* **243**, 60–69.

Schimke, R. T., Rhoads, R. E., Palacios, R., and Sullivan, D. (1973). Ovalbumin mRNA, complementary DNA and hormone regulation in chick oviduct. *Acta Endocrinol. (Copenhagen), Suppl.* **180**, 357–379.

Schneider, E. L., and Mitsui, Y. (1976). The relationship between *in vitro* cellular aging and *in vivo* human age. *Proc. Natl. Acad. Sci. U.S.A.* **73**, 3584–3588.

Schneider, J. A., and Weiss, M. C. (1971). Expression of differentiated functions in hepatoma cell hybrids. I. Tyrosine aminotransferase in hepatoma–fibroblast hybrids. *Proc. Natl. Acad. Sci. U.S.A.* **68**, 127–131.

Schubert, D., and Jacob, F. (1970). 5-bromodeoxyuridine-induced differentiation of a neuroblastoma. *Proc. Natl. Acad. Sci. U.S.A.* **67**, 247–254.

Schubert, D., Humphreys, S., Baroni, C., and Cohn, M. (1969). *In vitro* differentiation of a mouse neuroblastoma. *Proc. Natl. Acad. Sci. U.S.A.* **64**, 316–323.

Schwaber, J., and Cohen, E. (1973). Human–mouse somatic cell hybrid clone secreting immunoglobulins of both parental types. *Nature (London)* **244**, 444–445.

Schwaber, J., and Cohen, E. (1974). Pattern of immunoglobulin synthesis and assembly in a human–mouse somatic cell hybrid clone. *Proc. Natl. Acad. Sci. U.S.A.* **71**, 2203–2207.

Scriver, C. R., and Clow, C. L. (1980). Phenylketonuria: Epitome of human biochemical genetics. *N. Engl. J. Med.* **303**, 1336–1342.

See, Y. P., Carlsen, S. A., Till, J. E., and Ling, V. (1974). Increased drug permeability in Chinese hamster ovary cells in the presence of cyanide. *Biochim. Biophys. Acta* **373**, 242–252.

Seegmiller, J. E. (1976). Inherited deficiency of hypoxanthine guanine phosphoribosyl transferase in X-linked uric aciduria (the Lesch–Nyhan syndrome and its variants). *Adv. Hum. Genet.* **6**, 75–163.

Seegmiller, J. E., Rosenbaum, F. M., and Kelly, W. N. (1967). An enzyme defect associated with a sex-linked human neurological disorder. *Science* **155**, 1682.

Seidman, J., and Leder, P. (1978). Arrangement and rearrangement of antibody genes. *Nature (London)* **276**, 790–795.

Sekiguchi, T., and Sekiguchi, F. (1973). Interallelic complementation in hybrid cells derived from Chinese hamster diploid clones deficient in hypoxanthine–guanine phosphoribosyl-transferase activity. *Exp. Cell Res.* **77**, 391–403.

Shainberg, A., Yagil, G., and Yaffe, D. (1971). Alterations of enzymatic activities during muscle differentiation *in vitro*. *Dev. Biol.* **25**, 1–29.

Shani, M., Zevin-Sonkin, D., Saxwl, O., Carmon, U., Darcoff, D., Nudel, U., and Yaffe, D. (1981). The correlation between the synthesis of skeletal muscle actin, myosin heavy chain, and myosin light chain and the accumulation of corresponding mRNA sequences during myogenesis. *Dev. Biol.* **86**, 183–192.

Shapiro, L. J., Weiss, R., Buxman, M., Vidgoff, J., Dimond, R., Roller, J., and Wells, R. (1978). Enzymatic basis of typical X-linked ichthyosis. *Lancet* **2**, 756–757.

Shapiro, N. I., Knalizev, A. E., Luss, E. V. Marshak, M. I., Petrova, O. V., and Uarshaver, N. B. (1972). Mutagenesis in cultured mammalian cells. I. Spontaneous gene mutation in human and Chinese hamster cells. *Mutat. Res.* **15**, 203–214.

Sharma, S., Nirenberg, M., and Klee, W. (1975a). Morphine receptors as regulators of adenylate cyclase activity. *Proc. Natl. Acad. Sci. U.S.A.* **72**, 590–594.

Sharma, S., Klee, W., and Nirenberg, M. (1975b). Dual regulation of adenylate cyclase accounts for narcotic dependence and tolerance. *Proc. Natl. Acad. Sci. U.S.A.* **72**, 3092–3096.

Sharp, J. D., Capecchi, N. E., and Capecchi, M. R. (1973). Altered enzymes in drug resistant variants of mammalian tissue cultured cells. *Proc. Natl. Acad. Sci. U.S.A.* **70**, 3145–3149.

Shin, S. I. (1974). Nature of mutations conferring resistance to 8-azaguanine in mouse cell lines. *J. Cell Sci.* **14**, 235–251.

Shiraishi, Y., and Sandberg, A. (1978). Effects of mitomycin C on sister chromatid exchange in normal and Bloom's syndrome cells. *Mutat. Res.* **49**, 233–238.

Shows, T. B., and Brown, J. A. (1975). Human X-linked genes regionally mapped utilizing X-autosome translocations and somatic cell hybrid. *Proc. Natl. Acad. Sci. U.S.A.* **72**, 2125–2127.

Shuttlesworth, E. (1972). Barrier phenomena in brain tumors. *Prog. Exp. Tumor Res.* **17**, 279–290.

Sibatani, A. (1981). Molecular biology: A paradox, illusion and myth. *Trends Biol. Sci.* **6**(6), 4–9.

Sibley, C., and Tomkins, G. (1974). Isolation of lymphoma cell variants resistant to killing by glucocorticoids. *Cell* **2**, 213–220.

Siciliano, M. J., Siciliano, J., and Humphrey, R. M. (1978). Electrophoretic shift mutants in Chinese hamster ovary cells: Evidence for genetic diploidy. *Genetics* **75**, 1919–1923.

Simantov, R., and Sachs, L. (1975). Temperature sensitivity of cyclic adenosine $3' : 5'$-monophosphate-binding proteins and the regulation of growth and differentiation in neuroblastoma cells. *J. Biol. Chem.* **250**, 3236–3242.

Siminovitch, L. (1976). On the nature of heritable variation in cultured somatic cells. *Cell* **7**, 1–11.

Simons, J. W. (1974). Dose response relationships for mutants in mammalian somatic cells *in vitro*. *Mutat. Res.* **25**, 219–227.

Sinensky, M., Armagast, S., Mueller, G., and Torget, R. (1980). Somatic cell genetic analysis of regulation of expression of 3-hydroxy-3-methylglutaryl-coenzyme A reductase. *Proc. Natl. Acad. Sci. U.S.A.* **77**, 6621–6623.

Singer, S. J., and Nicolson, G. (1972). The fluid mosaic model of the structure of cell membranes. *Science* **175**, 720–731.

Singh, T., Roth, C., Gottesman, M., and Pastan, I. (1981). Characterization of cyclic AMP-resistant Chinese hamster ovary cell mutants lacking type I protein kinase. *J. Biol. Chem.* **256**, 926–932.

Siniscalo, M. (1979). Human gene mapping and cancer biology. *In* "Human Genetics: Possibilities and Realities" (R. Porter, ed.), pp. 283–327. Elsevier/North-Holland, New York.

Sleigh, M., Topp, W., Hanich, R., and Sambrook, J. (1978). Mutants of SV40 with an altered small t protein are reduced in their ability to transform cells. *Cell* **14**, 79–88.

Smith, H. H., Kao, K. N., and Combatti, N. C. (1976). Interspecific hybridization by protoplast fusion in *Nicotiana:* Confirmation and extension. *J. Hered.* **67**, 123–128.

Smith, J. R., and Hayflick, L. (1974). Variation in the life span of clones derived from human diploid cell strains. *J. Cell Biol.* **62**, 48–53.

Smith, M., and Hirschhorn, K. (1978). Location of the genes for human heavy chain immunoglobulin to chromosome 6. *Proc. Natl. Acad. Sci. U.S.A.* **75**, 3367–3371.

Soderberg, K. L., Ditta, G. S., and Scheffler, I. E. (1977). Mammalian cells with defective mitochondrial functions: A Chinese hamster mutant cell line lacking succinate dehydrogenase activity. *Cell* **10**, 697–702.

Somers, D. G., Pearson, M. L., and Ingles, C. J. (1975). Isolation and characterization of an alpha amanitin resistant myoblast mutant cell line possessing alpha amanitin resistant RNA polymerase II. *J. Biol. Chem.* **250**, 4825–4831.

Southern, E. M. (1975). Detection of specific sequences among DNA fragments separated by gel electrophoresis. *J. Mol. Biol.* **98**, 503–517.

Sparkes, R. S., and Weiss, M. C. (1973). Expression of differentiated functions in hepatoma cell hybrids: Alanine aminotransferase. *Proc. Natl. Acad. Sci. U.S.A.* **70**, 377–381.

Sparkes, R. S., Mohandas, T., Sparkes, M. C., and Shulkin, J. D. (1979). *p*-Uridyltransferase in Chinese hamster × human galactosemia somatic cell hybrids. *Biochem. Genet.* **17**, 683–692.

Springer, T., Galfre, G., Secher, D., and Milstein, C. (1978). Monoclonal xenogenic antibodies to murine cell surface antigens: Identification of novel leukocyte differentiation antigens. *Eur. J. Immunol.* **8**, 539–551.

Stamato, T. D., and Hochmann, L. (1975). A replica plating method for CHO cells using nylon cloth. *Cytogenet. Cell Genet.* **15,** 372–379.

Stanbridge, E. J., and Wilkinson, J. (1978). Analysis of malignancy in human cells: Malignant and transformed phenotypes are under separate genetic control. *Proc. Natl. Acad. Sci. U.S.A.* **75,** 1466–1469.

Stanley, P., Callibot, V., and Siminovitch, L. (1975a). Stable alterations at the cell membrane of Chinese hamster ovary cells resistant to the cytotoxicity of phytohemagglutinin. *Somatic Cell Genet.* **1,** 3–26.

Stanley, P., Callibot, V., and Siminovitch, L. (1975b). Selection and characterization of eight phenotypically distinct lines of lectin resistant Chinese hamster ovary cells. *Cell* **6,** 121–128.

Steglich, C., and DeMars, R. (1982). Mutations causing deficiency of APRT in fibroblasts cultured from human heterozygotes for mutant APRT alleles. *Somatic Cell Genet.* **8,** 115–141.

Steinberg, R. A., Levinson, B. B., and Tomkins, G. M. (1975). "Superinduction" of tyrosine aminotransferase by actinomycin D: A reevaluation. *Cell* **5,** 29–35.

Steinberg, R. A., O'Farrell, P. Friedrich, U., and Coffino, P. (1977). Mutations causing charge alterations in regulatory subunit of the cAMP dependent protein kinase of cultured S49 lymphoma cells. *Cell* **10,** 381–391.

Steinberg, R. A., van Daalen Wetters, T., and Coffino, P. (1978). Kinase-negative mutants of S49 mouse lymphoma cells carry a transdominant mutation affecting expression of cAMP-dependent protein kinase. *Cell* **15,** 1351–1361.

Stellwagen, R. H., and Tomkins, G. M. (1971a). Preferential inhibition by 5-bromodeoxyuridine of the synthesis of tyrosine aminotransferase in hepatoma cell cultures. *J. Mol. Biol.* **56,** 167–182.

Stellwagen, R. H., and Tomkins, G. M. (1971b). Differential effect of 5-bromodeoxyuridine on the concentrations of specific enzyme in hepatoma cells in cultures. *Proc. Natl. Acad. Sci. U.S.A.* **68,** 1147–1150.

Stenevi, U., Bjorklund, A., and Svendgaard, N. (1976). Transplanatation of central and peripheral monoamine neurons to the adult rat brain: Techniques and conditions for survival. *Brain Res.* **114,** 1–20.

Strehler, B. L. (1962). "Time, Cells and Aging." Academic Press, New York.

Strehler, B. L. (1979). Aging research: Current and future. *J. Invest. Dermatol.* **73,** 2–7.

Strom, C. M., and Dorfman, A. (1976). Amplification of moderately repetitive DNA sequences during chick cartilage differentiation. *Proc. Natl. Acad. Sci. U.S.A.* **73,** 3428–3432.

Strom, C. M., Moscona, M., and Dorfman, A. (1978). Amplification of DNA sequences during chicken cartilage and neural retina differentiation. *Proc. Natl. Acad. Sci. U.S.A.* **75,** 4451–4454.

Subak-Sharpe, H., Burk, R. P., and Pitts, J. D. (1969). Metabolic cooperation between biochemically marked mammalian cells in tissue culture. *J. Cell Sci.* **4,** 353–367.

Summers, W. P., and Handschumacher, R. E. (1973). The rate of mutation of L5178Y asparaginedependent mouse leukemia cells to asparagine independence and its biological consequences. *Cancer Res.* **33,** 1775–1779.

Sun, N. C., Chang, C. C., and Chu, E. H. (1974). Chromosome assignment of the human gene for galactose-1-phosphate uridyltransferase. *Proc. Natl. Acad. Sci. U.S.A.* **71,** 404–407.

Suttle, D. P., and Stark, G. R. (1979). Coordinate overproduction of orotate phosphoribosyltransferase and orotidine-5′-phosphate decarboxylase in hamster cells resistant to pyrazofurin and 6-azaguanine. *J. Biol. Chem.* **254,** 4602–4607.

Suzuki, F., Kashimoto, M., and Horikawa, M. (1971). A replica plating method of cultured mammalian cells for somatic cell genetics. *Exp. Cell Res.* **68,** 476–479.

Suzuki, N., and Okada, S. (1975). Location of the *ala32* gene replication in the cell cycle of cultured mammalian cells. L5178Y. *Mutat. Res.* **30,** 111–116.

Swan, D., D'Eustachio, P., Leinwand, L., Seidman, J., Keithley, D., and Ruddle, F. (1979). Chromosomal assignment of the mouse κ light chain genes. *Proc. Natl. Acad. Sci. U.S.A.* **76**, 2735–2739.

Szpirer, C. (1974). Reactivation of chick erythrocyte nuclei in heterokaryons with rat hepatoma cells. *Exp. Cell Res.* **83**, 47–54.

Szpirer, C., and Szpirer, J. (1976). Extinction, retention and induction of serum protein secretion in hepatoma–fibroblast hybrids. *Differentiation* **5**, 97–99.

Szpirer, J., Szpirer, C., and Wanson, J. (1980). Control of serum protein production in hepatocyte hybridomas: Immortalization and expression of normal hepatocyte genes. *Proc. Natl. Acad. Sci. U.S.A.* **77**, 6616–6620.

Szybalski, W., and Smith, M. (1959). Genetics of human cell lines. I. 8-azaguanine resistance, a selective single step marker. *Proc. Soc. Exp. Biol. Med.* **101**, 662–666.

Szybalski, W., and Szybalska, E. H. (1962). Drug sensitivity as a genetic marker for human cell lines. *Univ. Mich. Med. Bull.* **28**, 277–293.

Szybalski, W., Szybalska, E. H., and Ragni, G. (1962). Genetic studies with human cell lines. *Natl. Cancer Inst. Monogr.* **7**, 75–89.

Taylor, M. W., Pipkorn, J. H., Tokito, M. K., and Pozzattig, R. O., Jr. (1977). Purine mutants of mammalian cell lines. III. Control of purine biosynthesis in adenine phosphoribosyl transferase mutants of CHO cells. *Somatic Cell Genet.* **3**, 195–206.

Thacker, J., and Cox, R. (1975). Mutation induction and inactivation in mammalian cells exposed to ionizing radiation. *Nature (London)* **258**, 429–431.

Thacker, J., Stephens, M., and Stretch, A. (1976). Factors affecting the efficiency of purine analogues as selective agents for mutants of mammalian cells induced by ionizing radiation. *Mutat. Res.* **35**, 465–478.

Thomas, C. (1977). The fanciful future of gene transfer experiments. *In* "Genetic Interaction and Gene Transfer" (C. Anderson, ed.), pp. 348–360. Brookhaven Natl. Lab., Upton, New York.

Thompson, E. J. (1980). Tissue culture of dystrophic muscle cells. *Br. Med. Bull.* **36**, 181–185.

Thompson, E. J., and Gelehrter, T. (1971). Expression of tyrosine amino transferase activity in somatic cell heterokaryons: Evidence for negative control of enzyme expression. *Proc. Natl. Acad. Sci. U.S.A.* **68**, 2589–2593.

Thompson, L., Stanners, C., and Siminovitch, L. (1975). Selection by [3H] amino acids of CHO cell mutants with altered leucyl-transfer and asparagyl-transfer RNA synthesis. *Somatic Cell Genet.* **1**, 157–208.

Thompson, L., Rubin, J., Cleaver, J., Whitmore, G., and Brookman, K. (1980). A screening method for isolating DNA repair-deficient mutants of CHO cells. *Somatic Cell Genet.* **6**, 391–405.

Todaro, G., and Green, H. (1963). Quantitative studies of the growth of mouse embryo cells in culture and their development into established lines. *J. Cell Biol.* **17**, 299.

Tomkins, G., Gelehrter, T., Granner, D., Martin, D., Samuels, H., and Thompson, E. (1969). Control of specific gene expression in higher organisms. *Science* **166**, 1474–1480.

Tomkins, G. C., Stanbridge, E. J., and Hayflick, L. (1974). Viral probes of aging in the human diploid cell strain W1-38. *Proc. Soc. Exp. Biol. Med.* **146**, 385–393.

Tsutsui, T., Crawford, B., Ts'o, P., and Barrett, J. (1981). Comparison between mutagenesis in normal and transformed Syrian hamster fibroblasts. Differences in the temporal order of HPRT gene replication. *Mutat. Res.* **80**, 357–371.

Unakul, Q., Johnson, R., Rao, P., and Hsu, T. (1973). Giemsa banding in prematurely condensed chromosomes obtained by cell fusion. *Nature (London), New Biol.* **242**, 106–107.

Valbuena, O., Marcu, K., Croce, C., Huebner, K., Weigert, M., and Perry, R. B. (1978). Chromosomal location of mouse immunoglobulin genes. *Proc. Natl. Acad. Sci. U.S.A.* **75**, 2883–2887.

Van Diggelen, O. P., Donahue, T. F., and Seung-Il S. (1979). Basis for differential cellular sensitivity to 8-azaguanine and 6-thioguanine. *J. Cell. Physiol.* **98,** 59–72.

Van Zeeland, A. A., and Simons, J. W. (1975). Ploidy level and mutation to hypoxanthine–guanine phosphoribosyltransferase (HGRPT) deficiency in Chinese hamster cells. *Mut. Res.* **28,** 239–250.

Veomett, G., Prescott, D. M., Shay, J., and Porter, K. R. (1974). Reconstruction of mammalian cells from nuclear and cytoplasmic components separated by treatment with cytochalasin B. *Proc. Natl. Acad. Sci. U.S.A.* **71,** 1999–2002.

Veselý, J., and Čihák, A. (1973). Resistance of mammalian tumor cells toward pyrimidine analogues. *Oncology* **28,** 204–226.

Vogel, F., and Motulsky, A. G. (1980). "Human Genetics: Problems and Approaches." Springer-Verlag, Berlin and New York.

Wade, N. (1980). University and drug firm battle over billion-dollar gene. *Science* **209,** 1492–1494.

Wahl, G. M., Hughes, S. H., and Capecchi, M. R. (1974). Immunological characterization of HPRT mutants of mouse L cells. Evidence for mutations at different loci in the HPRT gene. *J. Cell. Physiol.* **85,** 307–320.

Wahl, G. M., Padgett, R. A., and Stark, G. R. (1979). Gene amplification causes overproduction of the first three enzymes of UMP synthesis in *N*-(phosphonacetyl)-L-aspartate-resistant hamster cells. *J. Biol. Chem.* **254,** 8679–8689.

Wallace, D. D., Bunn, C. L., and Eisenstadt, J. M. (1975). Cytoplasmic transfer of chloramphenicol resistance in human tissue culture cells. *J. Cell Biol.* **67,** 174–188.

Wang, R. (1976). Effect of room fluorescent light on the deterioration of tissue culture medium. *In Vitro* **12,** 19–22.

Watkins, J. F., and Grace, D. (1967). Studies on the surface antigens of interspecific mammalian cell heterokaryons. *J. Cell Sci.* **2,** 193–204.

Watson, B., Gormley, I. P., Gardiner, S. E., Evans, H. J., and Harris, H. (1972). Reappearance of murine hypoxanthine–guanine phosphoribosyltransferase activity in mouse A9 cells after attempted hybridization with human cell lines. *Exp. Cell Res.* **75,** 401–409.

Waymire, J., Prasad, K., and Weiner, N. (1972). Regulation of tyrosine hydroxylase activity in cultured mouse neuroblastoma cells elevation induced by analogs of adenosine 3'5' cyclic monophosphate. *Proc. Natl. Acad. Sci. U.S.A.* **69,** 2241–2245.

Weinstock, I., Jones, K. B., and Behrendt, J. R. (1978). Development of normal and dystrophic chick muscle in tissue culture. *J. Neuro. Sci.* **39,** 71–83.

Weiss, M. C., and Chaplain, M. (1971). Expression of differentiated functions in hepatoma cell hybrids: Reappearance of tyrosine aminotransferase inducibility after the loss of chromosomes. *Proc. Natl. Acad. Sci. U.S.A.* **68,** 3026–3030.

Weiss, M. C., Sparkes, R. S., and Bertolotti, R. (1975). Expression of differentiated functions in cell hybrids. IX. Extinction and reexpression of the liver-specific enzymes in rat hepatoma–Chinese hamster fibroblast hybrids. *Somatic Cell Genet.* **1,** 27–40.

Weiss, P. (1939). "Principles of Development." Holt, New York.

Westerveld, A., Visser, R., Freeke, M. A., and Bootsma, D. (1972). Evidence for linkage of PGK, HPRT, and G6PD in Chinese hamster cells studied by using a relationship between gene multiplicity and enzyme activity. *Biochem. Genet.* **7,** 33–40.

Whitlock, J. P., Gelboin, H. V., and Coon, H. G. (1976). Variation in aryl hydrocarbon (benzo[*a*]pyrene) hydroxylase activity in heteroploid and predominantly diploid rat liver cells in culture. *J. Cell Biol.* **70,** 217–225.

Wickens, M. P., Woo, S., O'Malley, B., and Gurdon, J. (1980). Expression of chicken chromosomal ovalbumin gene injected into frog oocyte nuclei. *Nature (London)* **285,** 628–634.

Widman, L., Golden, J., and Chasin, L. (1979). Immortalization of normal liver functions in cell culture rat hepatocyte hepatoma cell hybrids expressing ornithine carbamoyl transferase activity. *J. Cell. Physiol.* **100,** 391–400.

Wiener, F., Klein, G., and Harris, H. (1973). The analysis of malignancy by cell fusion. IV. Hybrids between tumor cells and a malignant L cell derivative. *J. Cell Sci.* **12**, 253–261.

Wigler, M., Pellicer, A., Silverstein, S., Axel, R., Urlaub, G., and Chasin, L. (1979a). DNA mediated transfer of the adenine phosphoribosyltransferase locus into mammalian cells. *Proc. Natl. Acad. Sci. U.S.A.* **76**, 1373–1376.

Wigler, M., Sweet, R., Sim, G., Wold, B., Pellicer, A., Lacy, E., Manitas, T., Silverstein, S., and Axel, R. (1979b). Transformation of mammalian cells with genes from prokaryotes and eukaryotes. *Cell* **16**, 777–785.

Wiktor, T., and Koprowski, H. (1978). Monoclonal antibodies against rabies virus produced by somatic cell hybridization: Detection of antigenic variants. *Proc. Natl. Acad. Sci. U.S.A.* **75**, 3938–3942.

Wild, O., and Hellkuhl, B. (1976). Isolation of mammalian cell mutants deficient in glucose-6-phosphate-dehydrogenase by means of a replica plating technique. *Hum. Genet.* **32**, 315–322.

Willecke, K., and Ruddle, F. H. (1975). Transfer of the human gene for hypoxanthine–guanine phosphoribosyltransferase via isolated human metaphase chromosomes into mouse L-cells. *Proc. Natl. Acad. Sci. U.S.A.* **72**, 1792–1796.

Willecke, K., Mierau, R., Kruger, A., and Lange, R. (1978). Chromosomal gene transfer of human cytosol thymidine kinase into mouse cells. *Mol. Gen. Genet.* **161**, 49–57.

Williams, A., Galfre, G., and Milstein, C. (1977). Analysis of cell surfaces by xenogenic myeloma-hybrid antibodies: Differentiation antigens of rat lymphocytes. *Cell* **12**, 663–643.

Williams, G. J., van der Horst, J., and Bootsma, D. (1977). Transfer of the human genes coding for thymidine kinase and galactokinase to Chinese hamster cells and human–Chinese hamster cell hybrids. *Somatic Cell Genet.* **3**, 281–293.

Willing, M., Nienhuis, A., and Anderson, W. F. (1979). Selective activation of human beta- but not alpha-globin gene in human fibroblast × mouse erythroleukaemia cell hybrids. *Nature (London)* **277**, 538.

Witkin, E. M. (1976). Ultraviolet mutagenesis and inducible DNA repair in *Escherichia coli*. *Bacteriol. Rev.* **40**, 869–907.

Witney, F. R., and Taylor, M. W. (1978). Role of adenine phosphoribosyltransferase in adenine uptake in wild type and APRT⁻-mutants of CHO. *Biochem. Genet.* **16**, 917–926.

Wohlhueter, R., and Plagemann, P. (1980). The roles of transport and phosphorylation in nutrient uptake in cultured animal cells. *Int. Rev. Cytol.* **64**, 171–240.

Wohlhueter, R., Marz, R., Graff, J., and Plagemann, P. (1979). A rapid mixing technique to measure transport in suspended animal cells: Application to nucleoside transport in Novikoff rat hepatoma cells. *Meth. Cell Biol.* **20**, 211–236.

Woo, S., Dugaiczyk, A., Tsai, M., Lai, E., Catterall, J., and O'Malley, B. (1978). The ovalbumin gene: Cloning and molecular organization of the entire natural gene. *Proc. Natl. Acad. Sci. U.S.A.* **76**, 2253–2257.

Worthy, T. E., Grobner, W., and Kelley, W. N. (1974). Hereditary orotic aciduria: Evidence for a structural gene mutation. *Proc. Natl. Acad. Sci. U.S.A.* **71**, 3031–3035.

Wright, E. D., Goldfarb, P. S. G., and Subak-Sharpe, J. H. (1976). Isolation of variant cells with defective metabolic cooperation (MEC⁻) from polyoma virus transformed Syrian hamster cells. *Exp. Cell Res.* **103**, 63–77.

Wright, W. E. (1978). The isolation of heterokaryons and hybrids by a selective system using irreversible biochemical inhibitors. *Exp. Cell Res. 112*, 395–407.

Wright, W. E., and Hayflick, L. (1975). Nuclear control of cellular aging demonstrated by hybridization of anucleate and whole cultured normal human fibroblasts. *Exp. Cell Res.* **96**, 113–121.

Wullems, G. J., van der Horst, J., and Bootsma, P. (1975). Incorporation of isolated chromosomes and induction of hypoxanthine phosphoribosyltransferase in Chinese hamster cells. *Somatic Cell Genet.* **1**, 137–152.

Wullems, G. J., van der Horst, J., and Bootsma, P. (1976a). Expression of human hypoxanthine phosphoribosyl transferase in China hamster cells treated with isolated human chromosomes. *Somatic Cell Genet.* **2**, 155–164.

Wullems, G. J., van der Horst, J., and Bootsma, D. (1976b). Transfer of the human X chromosome to human Chinese hamster cell hybrids via isolated HeLa metaphase chromosomes. *Somatic Cell Genet.* **2**, 359–371.

Wullems, G. J., van der Horst, J., Bootsma D. (1977). Transfer of the human genes coding for thymidine kinase and galactokinase to Chinese hamster cells and human–Chinese hamster cell hybrids. *Somatic Cell Genet.* **3**, 281–293.

Yaffe, D. (1968). Retention of differentiation potentialities during prolonged cultivation of myogenic cells. *Proc. Natl. Acad. Sci. U.S.A.* **68**, 477–483.

Yogeeswaran, G., Murray, R., Pearson, M., Sanwal, B., McMorris, F., and Ruddle, F. (1973). Glycosphingolipids of clonal lines of mouse neuroblastoma and neuroblastoma × L cell hybrids. *J. Biol. Chem.* **248**, 1231–1239.

Yotti, L. P., Chang, C. C., and Trosko, J. E. (1979). Elimination of metabolic cooperation in Chinese hamster cells by a tumor promoter. *Science* **206**, 1089–1091.

Zallin, R. J., and Montague, W. (1974). Changes in adenylate cyclase, cyclic AMP, and protein kinase levels in chick myoblasts and their relationship to differentiation. *Cell* **2**, 103–108.

Zelle, B., Reynolds, R., Kottenhagen, M., Schuite, A., and Lohman, P. (1980). The influence of the wavelength of ultraviolet radiation on survival, mutation induction and DNA repair in irradiated Chinese hamster cells. *Mutat. Res.* **72**, 491–509.

Zeuthen, J., Stenman, S., Fabricius, H. A., and Nilsson, K. (1976). Expression of immunoglobulin synthesis in human myeloma × nonlymphoid cell heterokaryons: Evidence for negative control. *Cell Differ.* **4**, 369–383.

Zylka, J. M., and Plagemann, P. G. (1975). Purine and pyrimidine transport by cultured Novikoff cells. Specifications and mechanisms of transport and relationship to phosphoribosylation. *J. Biol. Chem.* **250**, 5756–5767.

Index

CELL BIOLOGY: A Series of Monographs

EDITORS

D. E. BUETOW

Department of Physiology
and Biophysics
University of Illinois
Urbana, Illinois

I. L. CAMERON

Department of Anatomy
University of Texas
Health Science Center at San Antonio
San Antonio, Texas

G. M. PADILLA

Department of Physiology
Duke University Medical Center
Durham, North Carolina

A. M. ZIMMERMAN

Department of Zoology
University of Toronto
Toronto, Ontario, Canada

G. M. Padilla, G. L. Whitson, and I. L. Cameron (editors). THE CELL CYCLE: Gene-Enzyme Interactions, 1969

A. M. Zimmerman (editor). HIGH PRESSURE EFFECTS ON CELLULAR PROCESSES, 1970

I. L. Cameron and J. D. Thrasher (editors). CELLULAR AND MOLECULAR RENEWAL IN THE MAMMALIAN BODY, 1971

I. L. Cameron, G. M. Padilla, and A. M. Zimmerman (editors). DEVELOPMENTAL ASPECTS OF THE CELL CYCLE, 1971

P. F. Smith. The BIOLOGY OF MYCOPLASMAS, 1971

Gary L. Whitson (editor). CONCEPTS IN RADIATION CELL BIOLOGY, 1972

Donald L. Hill. THE BIOCHEMISTRY AND PHYSIOLOGY OF *TETRAHYMENA*, 1972

Kwang W. Jeon (editor). THE BIOLOGY OF AMOEBA, 1973

Dean F. Martin and George M. Padilla (editors). MARINE PHARMACOGNOSY: Action of Marine Biotoxins at the Cellular Level, 1973

Joseph A. Erwin (editor). LIPIDS AND BIOMEMBRANES OF EUKARYOTIC MICROORGANISMS, 1973

A. M. Zimmerman, G. M. Padilla, and I. L. Cameron (editors). DRUGS AND THE CELL CYCLE, 1973

Stuart Coward (editor). DEVELOPMENTAL REGULATION: Aspects of Cell Differentiation, 1973

I. L. Cameron and J. R. Jeter, Jr. (editors). ACIDIC PROTEINS OF THE NUCLEUS, 1974

Govindjee (editor). BIOENERGETICS OF PHOTOSYNTHESIS, 1975

James R. Jeter, Jr., Ivan L. Cameron, George M. Padilla, and Arthur M. Zimmerman (editors). CELL CYCLE REGULATION, 1978

Gary L. Whitson (editor). NUCLEAR–CYTOPLASMIC INTERACTIONS IN THE CELL CYCLE, 1980

Danton H. O'Day and Paul A. Horgen (editors). SEXUAL INTERACTIONS IN EUKARYOTIC MICROBES, 1981

Ivan L. Cameron and Thomas B. Pool (editors). THE TRANSFORMED CELL, 1981

Arthur M. Zimmerman and Arthur Forer (editors). MITOSIS/CYTOKINESIS, 1981

Ian R. Brown (editor). MOLECULAR APPROACHES TO NEUROBIOLOGY, 1982

Henry C. Aldrich and John W. Daniel (editors). CELL BIOLOGY OF *PHYSARUM* AND *DIDYMIUM*, Volume I: Organisms, Nucleus, and Cell Cycle, 1982; Volume II: Differentiation, Metabolism, and Methodology, 1982

John A. Heddle (editor). MUTAGENICITY: New Horizons in Genetic Toxicology, 1982

Potu N. Rao, Robert T. Johnson, and Karl Sperling (editors). PREMATURE CHROMOSOME CONDENSATION: Application in Basic, Clinical, and Mutation Research, 1982

George M. Padilla and Kenneth S. McCarty, Sr. (editors). GENETIC EXPRESSION IN THE CELL CYCLE, 1982

David S. McDevitt (editor). CELL BIOLOGY OF THE EYE, 1982

P. Michael Conn (editor). CELLULAR REGULATION OF SECRETION AND RELEASE, 1982

Govindjee (editor). PHOTOSYNTHESIS, Volume I: Energy Conversion by Plants and Bacteria, 1982; Volume II: Development, Carbon Metabolism, and Plant Productivity, 1982

John Morrow. EUKARYOTIC CELL GENETICS, 1983

In preparation

John F. Hartmann (editor). MECHANISM AND CONTROL OF ANIMAL FERTILIZATION, 1983

DATE DUE
REMINDER

JAN 03 '97

**Please do not remove
this date due slip.**